The Ultimate Terrorists

THE ULTIMATE TERRORISTS

Jessica Stern

HARVARD UNIVERSITY PRESS
Cambridge, Massachusetts
London, England
1999

Printed in the United States of America
Fourth printing, 2001

First Harvard University Press paperback edition, 2000

Library of Congress Cataloging-in-Publication Data

Stern, Jessica, 1958–
The ultimate terrorists / Jessica Stern.
p. cm.
Includes bibliographical references (p.) and index.
ISBN 0-674-61790-8 (cloth)
ISBN 0-674-00394-2 (paper)
1. Terrorism. 2. Terrorism—Technological innovations.
3. Weapons of mass destruction. I. Title
HV6431.S74 1999
303.6′25—dc21 98-42453

For Jeff

Contents

The Ultimate Terrorists

Terrorism Today

What if terrorists exploded a homemade nuclear bomb at the Empire State Building in New York City? A one-kiloton nuclear device—tiny by superpower standards—would ignite a fireball 300 feet in diameter that would demolish the Empire State Building and the 20,000 people who work there, leaving in their place a crater 120 feet wide. Much of the building, and everyone in it, would be vaporized by the intense heat. A shock wave would spread out from the blast site, exposing everything in its path to pressure as high as thousands of pounds per square inch. Components of the Empire State Building that had not vaporized would create a storm of concrete, glass, and steel missiles, which would be propelled thousands of feet by strong horizontal winds.

Buildings within 600 feet would collapse, as would the underground infrastructure of subways, wiring, and pipes. Gas mains would rupture, causing widespread fires. A bright light, many times brighter than the sun in the desert at noon, would be visible from neighboring states. People up to a quarter of a mile away would be killed or maimed as their clothing burst into flames from the heat. Those wearing dark suits, as New Yorkers are wont to do, would be particularly susceptible, since dark colors (and synthetics) absorb radiation. Radiation would quickly kill those within half a mile of the blast.

An updraft would suck up dirt and debris, including the crushed remains of the Empire State Building, into a radioactive mushroom cloud 10,000 feet high and the color of blood, tinted by red and brown nitrous acids and oxides of nitrogen. As it cooled the cloud would turn the white of ordinary cumulus. This grim marker would be visible for miles around.

In the first twenty-four hours radioactive particles ranging in size from fine powder to marbles—and even larger close to the burst point—would descend from the sky. A lethal dose would be delivered to anyone within an area a quarter-mile wide and nine miles long—as far north as the George Washington Bridge, or, depending on the wind, out into New Jersey and Brooklyn. Victims of this "early" fallout would die within two weeks. People as far as eighteen miles away would suffer radiation sickness. Even hundreds of miles downwind, cancer rates would rise, and long-lived isotopes would contaminate the area for years. Small radioactive particles would eventually be deposited over much of the earth. Because the bomb would explode close to the ground, the effects of fallout would be far more severe than at Hiroshima or Nagasaki.

How many people would die? It's hard to say, but the fallout alone might kill up to 100,000—in addition to those killed by blast, heat, or initial radiation close to the explosion. The death toll might easily reach twice that.[1]

But the ramifications for survivors of a nuclear explosion in America would go far beyond the shock and grief, funerals and fallout. Even a tiny nuclear detonation, hundreds or even thousands of times smaller than most of the bombs stockpiled by Russia and the United States, might permanently alter America's cherished balance between civil liberties and public safety. Even an attack with a conventional bomb might have this effect: after the bombing of the federal building in Oklahoma City many Americans jumped to the conclusion—without evidence—that it had been perpetrated by Middle Eastern terrorists. Arab Americans were subjected to harassment and prejudicial treatment. Four men of Middle Eastern origin were detained after the attack; one of them, a Jordanian American, was forcibly returned to the United States from London.[2]

A nuclear attack, on the Empire State Building or elsewhere, might evoke a far more extreme reaction. Surviving leaders might feel compelled to reinforce the government's authority to search out and deport foreign terrorists, and might call for measures that would violate civil rights. Phones might be tapped, foreigners' movements monitored. Mosques, for example, might be targeted for surveillance. Rights of free speech and free assembly might be curtailed. There would probably be much more sympathy for FBI snooping, for CIA spying, and for chips that monitor electronic conversations. Citizens might demand an expansion of the military's role in protecting civilians at home.[3] Within days, the American way of life might change substantially.

Even if the terrorists' homemade bomb failed to reach nuclear yield, it would nonetheless spread radioactive contamination, with devastating economic and psychological consequences. Few people would die from the bomb itself. But fear of radiation might cause panic, which could lead to deaths. For example, the radiation would not be high enough to require evacuation, but people might panic and try to leave the city, creating massive traffic jams and possibly accidents.

The U.S. Department of Energy predicts that such a bomb would cause very few deaths from cancer. But the economic and psychological costs would be formidable. If a bomb with some six pounds of plutonium exploded in Washington, D.C., 45,000 people might have to stay indoors for an undefined period afterward to avoid being exposed to fallout. And the public's fear of radiation would probably require that authorities clean up an area of about seventy-three square miles. Buildings would have to be scrubbed, topsoil removed, pavement hosed down. The cost would be likely to exceed $100 billion—around a third of the yearly defense budget for the United States.[4]

A terrorist attack using chemical or biological weapons would be far easier to accomplish, and could be equally devastating to public confidence and civil liberties. In recent years terrorists have been acquiring crude chemical and biological agents, and some have plotted or threatened to use them. It would be relatively easy to use them to poison agricultural commodities, infect livestock, or gas passengers on trains or planes.

Biological weapons have the potential to be as deadly as nuclear bombs. For example, 100 kilograms of anthrax, less than the amount

Iraq has produced, could kill up to 3 million people if dispersed under optimal conditions. In comparison, a Hiroshima-type fission bomb with a yield of 12,500 tons of TNT could kill up to 80,000, while a more powerful hydrogen bomb, with a yield of a million tons of TNT, could kill between 600,000 and 2 million. Few if any terrorists are likely to be capable of such attacks, but fear of poisons could make even a small-scale, low-technology incident—the kind terrorists are more likely to achieve—psychologically devastating.[5]

Unlike the effects of conventional or nuclear weapons, however, those of some biological and chemical agents can be reversed. If authorities were aware that an attack had taken place, victims could be treated. For some contagious agents, vaccines might prevent person-to-person spread. But if doctors did not know there had been a biological attack, they might think the victims had the flu, since the early symptoms are similar. By the time victims displayed unmistakable symptoms—perhaps days after the attack—it would be too late to save their lives. In the meantime, if the agent was contagious, thousands of others might be infected by coming into contact with the original victims.

Vulnerable Societies

Successful terrorists will choose their technology to exploit the vulnerabilities of a particular society. Modern societies are particularly susceptible to weapons that are capable of killing many people at one time—weapons of mass destruction (WMD). Their citizens tend to live, work, and travel in close proximity, providing concentrated targets. For the Aum Shinrikiyo cult in Japan, which released poison gas in Tokyo subway cars, gas was an effective weapon not only because of its capacity to inspire fear but also because in Tokyo many people go about their lives close together, especially on the subway.

Despite this concentration of population, many individuals in modern societies are isolated from one another. This creates fertile ground for the breeding of extremists and makes it possible for extremist groups to operate unnoticed. Again the Aum Shinrikiyo case is an example: the Japanese government's failure to detect the cult's plans suggests the possibility that terrorists may be able to pursue other WMD, even crude nuclear weapons, without being caught.

Americans face an additional vulnerability. First Amendment protections make it legal to advocate using WMD, to disseminate detailed instructions for producing them, and to advise would-be terrorists about how to evade detection.[6] The same Bill of Rights that makes Americans uniquely free also makes terrorism harder to combat. It prevents the government from banning poisoning manuals and makes it difficult for government agents to infiltrate and monitor terrorist groups. The Internet provides an easy way for terrorists to spread information around the world, to recruit, and to plan operations in secret. And Americans' fundamental wariness of government makes it easier to terrorize them; citizens have little faith in the government's ability to minimize fatalities in case of an attack. People who don't trust their government to protect them are more susceptible to panic, and panic may lead to loss of life.

Trends in Terrorism

Western societies enjoy extraordinary military and economic power today. But the "clash of civilizations" between "the West and the rest" predicted by Samuel Huntington is unlikely to take place exclusively—or even principally—on the battlefield. Some potential adversaries, Secretary of Defense William Cohen warned in 1997, believe that their only way to fight America, given U.S. military superiority, is to use WMD against U.S. troops or civilians. Violent Islamic extremists have already recognized that they cannot defeat the United States in a conventional war, but that they can impose significant pain through acts of terrorism. Right-wing extremists and other domestic groups are also likely to participate in this "clash."[7]

Islamic fundamentalists, Huntington observes, do not subscribe to the western belief in separation of church and state. Nor do they value the Enlightenment ideals of liberalism, human rights, equality, and the rule of law. The "rest" that oppose the West are inspired by radically different views of the relative importance of rights and responsibilities, liberty and authority, and equality and hierarchy, and of the relations between the individual and the group, the citizen and the state, and husband and wife.[8] The observations apply equally to home-grown extremist groups such as Christian Patriots (antigovernment white supremacists who subscribe to a doctrine known as Christian Identity, believing that blacks are subhuman and that Jews are the descendants of Satan).

During the 1970s, 8,114 terrorist incidents were reported around the world, resulting in 4,798 deaths and 6,902 injuries. During the 1980s the number of incidents increased nearly fourfold, to 31,426, with 70,859 deaths and 47,849 injuries. From 1990 to 1996 there were 27,087 incidents, causing 51,797 deaths and 58,814 injuries. The number of deaths due to acts of terrorism varies from year to year, but there is a clearly increasing trend. Between 1970 and 1995, on average, each year brought 206 more incidents and 441 more fatalities.[9] In 1996 the number of international incidents declined, but, according to the State Department, deaths and injuries continued to increase.

The 1990s were marked by several positive developments, along with some negative ones. The collapse of Communism dealt a severe blow to traditional left-wing terrorist groups: they are now bereft of sponsors, safe havens, and training camps in Europe. Countries such as Cuba, North Korea, and Libya, formerly engaged in supplying terrorists with training, weapons, or funds, have reportedly renounced this role. And there are encouraging signs that Palestinian groups and the IRA are choosing political alternatives to terrorist violence.[10]

On the negative side, Iran remains deeply involved in acts of terrorism committed by its own agents or by surrogate groups. Iran and Syria may have been involved in a June 1996 bombing of a U.S. military housing complex in Dhahran, Saudi Arabia, which killed nineteen Americans. Egypt has accused Iran of backing the Egyptian Islamic militants who attempted to assassinate President Mubarak in 1995. Sudan is also suspected in connection with this incident. Sudan provides paramilitary training as well as refuge for a number of extremely violent Middle Eastern groups, including the Abu Nidal organization, Lebanese Hezbollah, Hamas, and Gama'at al-Islamiyya. A Sudanese national pleaded guilty to complicity in a foiled plot to bomb the

United Nations and other New York City targets and claimed that a Sudanese official had offered to provide access to the U.N. building. Afghanistan currently hosts Osama bin Laden, a wealthy Saudi alleged to have financed the attack and believed to provide support for a broad group of violent Middle Eastern extremists. Several Middle Eastern terrorist groups have supporters inside the United States who might abet acts of terrorism on U.S. territory.[11]

The emergence of ad hoc radical fundamentalist Islamic groups, such as the multinational group involved in the bombing of the World Trade Center, is a particularly troubling development. These groups operate on a global scale and claim to act for Islam. According to the U.S. State Department, they have sources of funding around the world and are knowledgeable about modern explosives and weapons. Ad hoc groups can form quickly, need no headquarters, and have no recognized leaders; these characteristics make them more difficult to track and apprehend than members of established groups.[12]

Terrorist groups motivated by religious concerns are becoming more common. Of eleven international terrorist groups identified by the Rand Corporation in 1968, none was classified as religiously motivated. By 1994 a third of the forty-nine international groups identified were classified as religious. Religious groups are more likely than others to turn to WMD.[13]

The United States may have contributed to terrorist violence by training and financing the Mujahedeen in Afghanistan's war with the Soviet Union in the 1980s, leaving "thousands of highly trained Islamic militants who dispersed, [taking] with them an ideology of violence and revolution." One expert on Islamic extremism warns that "the mercenary groups of terrorism," under the influence of religious zealots in Iraq, Algeria, Libya, Sudan, and Iran, are "becoming more fanatical . . . The mullahs are convinced they are suffering from a situation for which they are not responsible. All the failures they made avoiding enlightenment, by avoiding technological revolution, they blame on the West. They will go to war because of this."[14]

Meanwhile Christian Patriots are growing in number. They are also showing signs of interest in biological weapons. Survivalists and white supremacists were implicated in three separate cases involving biological agents in 1995. In March two members of the Minnesota Patriots Council were arrested for producing ricin with which to assassinate a deputy U.S. marshal who had served papers on one of them for tax violations. In May, just six weeks after the Aum Shinrikiyo incident, Larry Wayne Harris, a former member of neo-Nazi organizations, bought three vials of *Yersinia pestis*, the bacterium that causes bubonic plague, which killed nearly a quarter of Europe's population in the mid-fourteenth century. Harris had ordered the bacteria from the American Type Culture Collection, the same organization that sold biological agents to Iraq. No law prohibited Harris or any other American from acquiring the agent. But Harris had misrepresented himself in his purchase order, and he was convicted of mail fraud. In December a survivalist was arrested for trying to carry 130 grams of ricin across the border into Canada. Agents who searched his house found castor beans, from which ricin is extracted, and three manuals on poisons.[15]

Why Now?

Five interrelated developments have increased the risk that terrorists will use nuclear, chemical, or biological weapons against civilian targets. First, such weapons are especially valuable to terrorists seeking to conjure a sense of divine retribution, to display scientific prowess, to kill large numbers of people, to invoke dread, or to retaliate against states that have used these weapons in the past. Terrorists motivated by goals like these rather than traditional political objectives are increasing in number.

Second, terrorists' motivations are changing. A new breed of terrorists—including ad hoc groups motivated by religious conviction or revenge, violent right-wing extremists, and apocalyptic and millenarian cults—appears more likely than the terrorists of the past to commit acts of extreme violence. Religious groups are becoming more common, and they are more violent than secular groups. Religious groups committed only 25 percent of the international terrorist incidents recorded in the Rand–St. Andrews Chronology in 1995, but they were responsible for 58 percent of the deaths. George Tenet, then Acting Director of the CIA, warned in 1997 that "fanatical" terrorists pose an "unprecedented threat" to the United States, and that a growing number of groups are investigating the feasibility of chemical, biological, and radiological weapons.[16]

Third, with the breakup of the Soviet Union, the black market now offers weapons, components, and knowhow. The Soviet nuclear-security system was designed during the Cold War to prevent Americans from stealing secrets, not to prevent theft by insiders. And that inadequate system has largely broken down. Hundreds of tons of nuclear material, the essential ingredient of nuclear weapons, are stored at vulnerable sites throughout the former Soviet Union, guarded only by underpaid, hungry, and disheartened people. Some of it is stored in gym-type lockers, secured with the equivalent of bicycle locks. At least eight thefts of materials that could be used to make nuclear weapons have been confirmed. Worse yet, the weapons themselves may be vulnerable to theft or unauthorized launch. While Russia's 6,000 long-range strategic weapons are protected by locks, making it impossible—at least in principle—to launch them without high-level authority, thousands of smaller tactical weapons have less sophisticated protection or no locks at all, making them both easier to steal and easier to detonate. Since many of Russia's nuclear custodians and weapons scientists are now unpaid or unemployed, they may eventually give in to financial pressures by selling their expertise or their wares abroad (see Chapter 6).

Fourth, chemical and biological weapons are proliferating, even in states known to sponsor terrorism. Some governments, including China, Russia, and North Korea, are exporting equipment that, while ostensibly intended for benign purposes, could be used to manufacture WMD. Iraq is an example of a state known to sponsor terrorism (see Chapter 7) that is developing chemical and biological weapons. Iraq has also used chemical weapons in acts of state terror against its own citizens. Despite its military defeat in 1991 and its commitment to destroy its WMD under U.N. inspection, Baghdad has allegedly threatened to use WMD against Britain and other nations, and it has the potential to do so. The situation in Iraq shows how difficult it is to prevent the proliferation of WMD. Preventive war did little to root out Iraq's weapons of mass destruction, and the most intrusive inspection regime ever devised has left inspectors guessing, especially about Iraq's biological weapons program.

Fifth, advances in technology have made terrorism with weapons of mass destruction easier to carry out. For example, the Internet allows terrorists to recruit from a larger pool of potential sympathizers and to communicate instantaneously. Advanced fermenting equipment makes it easier to optimize the growth of biological organisms, and new technologies for coating microorganisms make dissemination less difficult. (But technology also makes terrorism easier to thwart; see Chapter 8.)

Despite these developments, terrorists' use of weapons of mass destruction is likely to remain rare. Few terrorists will be capable of using these weapons except in small-scale incidents, and few will want to kill tens or hundreds of thousands of people. It is useful to think of the danger in terms of the concept of expected cost: the product of the probability of an event and its consequences. While the probability of WMD terrorism is low, its expected cost—in lives lost and in threats to civil liberties—is potentially devastating. Government officials will be remiss—and will be blamed—if they do not take measures to reduce the likelihood and severity of the threat.

Definitions

> The killing of soldiers and nearby civilians [is] to be defended only insofar as [it is] the product of a single intention, directed at the first and not the second. The argument suggests the great importance of taking aim in wartime, and it correctly restricts the targets at which one can aim.
>
> —Michael Walzer, 1977

Hundreds of definitions of terrorism are offered in the literature. Some focus on the perpetrators, others on their purposes, and still others on their techniques. But only two characteristics are critical for distinguishing terrorism from other forms of violence. First, terrorism is aimed at noncombatants. This is what makes it different from fighting in war. Second, terrorists use violence for a dramatic purpose: usually to instill fear in the targeted population. This deliberate evocation of dread is what sets terrorism apart from simple murder or assault.[1]

I define terrorism as an act or threat of violence against noncombatants with the objective of exacting revenge, intimidating, or otherwise influencing an audience. This definition avoids limiting perpetrator or purpose. It allows for a range of possible actors (states or their surrogates, international groups, or a single individual), for all putative goals (political, religious, or economic), and for murder for its own sake.[2]

WMD terrorism involves the most modern—and the most extreme—forms of random violence. Nuclear, chemical, and biological weapons are *inherently terrifying:* in most scenarios for their use, the fear they would cause would dwarf the injury and death. Dread of these weapons creates its own dangers: if victims panic and attempt to flee, they may spread contamination and disease still further.

These weapons are also *inherently indiscriminate*. Conventional weapons can be used either discriminately, to harm only soldiers, or indiscriminately, to harm noncombatants. While in principle chemical weapons can be exclusively used on the battlefield, and nuclear weapons can be used in counter-force strikes to target the enemy's weapons systems, in practice WMD harm noncombatants. It is impossible to aim at a particular target; only the most sophisticated militaries can use these weapons in open areas without putting noncombatants at risk.[3]

The effects of these weapons are also *inherently random*. The radius of injury depends on conditions that are impossible to control or to predict with certainty. The movement of aerosols, the virulence of microorganisms, the susceptibility of victims, and the spread of fallout all depend on exogenous variables like meteorological conditions and terrain. These weapons' fear-inspiring, all-encompassing, unpredictable nature is what makes them consummate instruments of terror (see Chapter 3).

Defining terrorism is more than an academic exercise. The definition inevitably determines the kind of data we collect and analyze, which in turn influences our understanding of trends and our predictions about the future. For example, the U.S. State Department analyzes only international terrorist incidents, that is, incidents involving citizens of more than one country. And the definition employed by the U.S. government confines terrorism to politically motivated violence. Although the FBI has recently started collecting data on domestic incidents in the United States, no government organization collects and analyzes data on all terrorism around the world. The lack of comprehensive data makes it impossible to analyze broad trends in terrorism, or to use empirical evidence to predict what kinds of domestic terrorists are most likely to be attracted to weapons of mass destruction.

How we define terrorism profoundly influences how we respond to it. If terrorism is always a crime (as distinct from war), then the Justice Department and the police are responsible for combating it, and it is legally difficult to call on the military in incidents on U.S. territory, even in situations (such as those involving chemical or biological agents) to which only the military is trained to respond. And yet if terrorists did employ such weapons on a massive scale they might kill as many Americans as would be killed in a "major regional conflict," the kind of war the Pentagon is prepared to fight. In the Gulf War, for example, 111 Americans died from enemy fire; hundreds of thousands might be killed in a terrorist incident using nuclear or biological weapons.[4]

Definitions of terrorism have changed with the times and the political environment. The word *terrorism* originated in revolutionary France, and the first definition in the *Oxford English Dictionary* refers to "government by intimidation as carried out by the party in power in France during the Revolution of 1789–1797." Cold War definitions tended to focus on surrogate warfare, since the West suspected the Warsaw Pact nations of training terrorists and instigating acts of terrorism. In 1989 Yonah Alexander defined terrorism as "a process of deliberate employment of psychological intimidation and physical violence by sovereign states and sub-national groups to attain strategic and political objectives in violation of law." Title 22 of the United States Code defines terrorism as "premeditated, politically motivated violence perpetrated against noncombatant targets by sub-national or clandestine agents." The U.S. Department of State appends a final phrase: "usually intended to influence a target audience." Echoing this idea of a target audience, the *OED*'s second definition is "a policy intended to strike with terror those against whom it is adopted; the employment of intimidation."[5] In the 1990s, with domestic extremists and religious cults becoming increasingly common and increasingly violent, a useful definition must encompass both international and domestic groups, and must include all ostensible motivations, including religious ones.

The characteristics of terrorism, as listed earlier, raise additional questions. How do we define "noncombatants"?[6] A soldier on a battlefield is unquestionably a combatant, but what if the soldier's country is not at war, and he is sleeping in a military housing complex, as nineteen U.S. soldiers were when they were killed by a bomb in Dhahran, Saudi Arabia, in June 1996? What if he is riding a bus that also carries civilians, as an Israeli soldier was when a suicide bomber attacked in Gaza in April 1995? Under these circumstances some would claim that the soldier should not be a military target. But what if troops are sent into a country engulfed by civil war? And what if the warring factions perceive those "peacekeeping" troops to be partisan? This question is likely to arise whenever the U.S. military involves itself in peacekeeping operations in other countries.

Another thorny issue is whether states themselves can be the perpetrators of terrorist acts. The answer is yes. States and their leaders can and do unleash terrorist violence against their own civilians, as Saddam Hussein did with chemical weapons against Iraqi Kurds; as Stalin did in acts of random violence against Soviet citizens; and as the Guatemalan government did for nearly forty years against its own people.

States have also used terrorism as an instrument of war, by deliberately attacking civilians in the hope of crushing enemy morale. In late 1940 the British Chiefs of Staff determined that Germany's morale was more vulnerable than its industry, and decided to bomb the centers of key German cities. Historians estimate that these attacks killed some 300,000 Germans, most of them civilians, and seriously injured 780,000 more. In Dresden alone, in the spring of 1945, when the war was virtually won, the bombing killed nearly 100,000 people.

The U.S. Army clearly intended to terrorize Japan into submission when it dropped atomic bombs on Hiroshima and Nagasaki. General Marshall claimed it was not enough to kill 100,000 Japanese in a single fire-bombing raid: "There were only two ways to win the war—either by going in after them or by shock. The atomic bomb was the shock action." Some planners wanted to drop the bomb in the Sea of Japan, he said: "Others wanted to drop it in a rice paddy to save the lives of the Japanese. But we only had two [atomic bombs], and the situation demanded shock action." Targeting civilians was considered necessary to maximize terror.[7]

Some ethicists condemn the terror bombing, even though President Truman authorized the use of atomic bombs as a way to shorten the war. A similar debate arises in the context of more conventional acts of terrorism not perpetrated by states.

Historical Context

The concept of terrorism has evolved over the centuries. Its early roots are in acts of assassination, regicide, and tyrannicide. Three historical terrorist organizations are of particular interest here: the Zealots-Sicarii, the Assassins, and the Thugs. All were inspired by religious conviction, and all were active internationally. Their weapons were primitive—the sword, the dagger, and the noose—and yet they were more destructive than any modern group.[8]

The Zealots-Sicarii, a Jewish group active in the first century CE, profoundly influenced the history of the Jews. They began by murdering individual victims with daggers and swords, and later they turned to open warfare. Their objective was to create a mass uprising against the Greeks in Judea and against the Romans, who governed both Greeks and Jews. The revolt had unforeseen and devastating consequences, leading to the destruction of the Jewish Temple and to the mass suicide at Masada. Later revolts inspired by the Zealots-Sicarii led to the extermination of the Jews in Egypt and Cyprus, the virtual depopulation of Judea, and the Exile, which remained central features of the Jewish experience for the next two thousand years. "It would be difficult to find terrorist activity in any historical period which influenced the life of a community more decisively," David Rapoport observes[9]—though the influence was not what the terrorists intended.

The Assassins, or Ismailis-Nizari, operated from 1090 to 1275. Their objective, like that of some of today's violent Islamist extremists, was to spread a purified version of Islam. Their technique was to stab their victims in broad daylight, which made the assailants' escape all but impossible. They considered their own lives to be sacrificial offerings. Their targets were prominent politicians or religious leaders who had refused to accept the new preaching. The Assassins seriously threatened the governments of several nations, including the Turkish Seljuk Empire in Persia and Syria.

The Thugs were active in India for at least six hundred years, until the late nineteenth century. No one knows precisely how many people they killed, but estimates range above 500,000. Their victims were travelers, whom they strangled, dismembered, buried, and robbed. The

Thugs' principal objective was to provide sacrificial offerings to the goddess Kali. They strove to prolong the death agony of their victims, believing Kali enjoyed the victims' terror.

In the late nineteenth century Europe was plagued by a series of sensational assassinations. Anarchists and social revolutionaries attacked kings, queens, members of the aristocracy, and government officials, creating a sense of high anxiety among the ruling elite. In 1878 terrorists attacked the emperor of Germany, the king of Spain, and the king of Italy. During the 1890s, known as the decade of the bomb, chiefs of state were murdered at the rate of nearly one per year. President Carnot of France was assassinated in 1894, the premier of Spain in 1897, the empress of Austria in 1898, the king of Italy in 1900, and the president of the United States in 1901. Many attempts were also made on the lives of lower-level officials throughout Europe.[10]

In Russia at the turn of the century the intellectual elite considered terrorist violence the only effective way to modernize Russian society. The acquittal of Vera Zasulich, who had attempted to kill the governor of Saint Petersburg, reflects this view, as does Turgenev's prose poem "The Threshold," a romantic portrayal of a young female terrorist, written shortly after Zasulich's trial. Despite incontrovertible evidence, the jury found Zasulich innocent because her motives—to protest ill treatment of political prisoners—were considered humane. This "miscarriage of justice," according to Richard Pipes, had profound implications, in that it gave authorities license to dispose of political cases outside the courts. During this "heroic" period of Russian terrorism, militant members of the populist group Zemlya I Volya (Land and Freedom) founded the terrorist organization Narodnaya Volya (The People's Will), which was responsible for numerous political assassinations between 1879 and 1881, culminating in the murder of Tsar Alexander II in 1881. Terrorist violence in Russia continued to worsen, setting the stage for the emergence of the police state. The terrorism of the late nineteenth and early twentieth centuries set a standard of brutality that became the modus operandi of the final two reigns of the Romanovs and of the Soviet regime.[11]

Although terrorism came later to the United States than to Europe, pre-twentieth-century America was not immune to such violence. The Civil War inspired terrorism on both sides. Resistant southerners formed an organization called the Ku Klux Klan to fight Reconstruction. Anarchists were active in the 1880s, especially in Chicago, where tens of thousands subscribed to periodicals calling for the violent overthrow of the state. When President McKinley was assassinated by an alleged anarchist militant, the government cracked down on anarchists, even rejecting would-be immigrants who admitted to opposing organized government.[12]

A new kind of terrorism emerged in the second half of the twentieth century: random violence. Especially since the 1970s, terrorists have increasingly targeted people who had no obvious connection to the terrorists' grievances. The late twentieth century also saw a resurgence of holy terror—the kind practiced by the Zealots-Sicarii, the Assassins, and the Thugs. Religiously inspired groups, as we shall see in Chapter 5, tend to be more violent than their secular counterparts.

Random violence, whether inspired by secular or religious concerns, raises the most serious ethical questions. It can be argued that government officials—especially in oppressive regimes—are not noncombatants. It can also be argued that although assassinating tyrants is a crime, it is sometimes morally justified. Thomas Aquinas argued that because tyrannical governments are directed toward the private good of the ruler, rather than the common good of the governed, the overthrowing of such governments is to be praised, unless it produces such disorder that the suffering of the governed is increased.[13] Aristotle and Cicero, among others, made similar arguments. But who has the right to identify a tyrant?

Morality

While terrorism is not war, the two raise similar ethical questions. Charles Francis Adams contrasted two views in 1903: "On the one side, it is contended that warfare should be strictly confined to combatants, and its horrors and devastation brought within the narrowest limits . . . But, on the other hand, it is insisted that such a method of procedure is mere cruelty in disguise; that war at best is Hell, and that true humanity lies in exaggerating that Hell to such an extent as to make it unendurable. By so doing, it is forced to a speedy end." The latter alternative is known as the realist view. Its proponents include General Sherman, who defended the burning of Atlanta in the Civil War by declaring, "War is cruelty and you cannot refine it."[14]

An alternative view is that war is a rule-governed activity; that it is, in Michael Walzer's words, a "moral world in the midst of hell"; that wars can be fought for just ends and carried out with just means. There are two separate components to the concept of just war (which philosophers call the just war tradition): the rationale for initiating the war (war's ends) and the method of warfare (war's means). Criteria for whether a war is just are divided into *jus ad bellum* (justice of war) and *jus in bello* (justice in war) criteria. Under the former, war is permissible when there are no better means for securing the peace, provided the following conditions are met: just cause, right authority, right intention, and proportionality. "In jus ad bellum terms," James Johnson explains, "the aim of the idea of proportionality is to ensure that the overall damage to human values that will result from the resort to force will be at least balanced evenly by the degree to which the same or other important values are preserved or protected."[15]

The *jus in bello* (justice in war) criteria concern adversaries' conduct rather than their ends. *Jus in bello* requires that the belligerents use methods proportional to their ends, and that they not directly target noncombatants.[16] This requirement is modified to allow for the "double effects" of acts of war that inadvertently kill civilians (what military strategists call collateral damage). What matters here, according to the philosopher Steven Lee, is intention: there is a morally relevant difference "between merely foreseeing the deaths of noncombatants as an effect of military activity and intending to bring about those deaths; the principle of discrimination rules out the activity only in the latter case."[17]

It is possible to be justified in initiating a war but to conduct it unjustly, and equally possible to initiate an unjust war but still conform to the law of war. The just war tradition requires that war meet both sets of criteria: both its ends (*jus ad bellum*) and its means (*jus in bello*) must be just. Such wars are rare indeed. Modern terrorism, while it may subjectively meet the first requirement, invariably fails the second, because terrorism deliberately targets noncombatants. But not all of the major moral traditions would proscribe terrorism, or the targeting of noncombatants, in all circumstances. Deontological ethical systems, such as Kant's, would do so, because they hold that the value of an act is intrinsic. The act is either good or evil in and of itself. So killing the innocent is wrong, regardless of its ultimate consequences.[18]

However, for those who subscribe to the moral tradition of consequentialism, only the consequences of an act count; its intrinsic value is irrelevant. Consequentialists believe the right act in any situation is the one that will produce the best overall outcome: if the good that comes to society as a whole exceeds the harm done to the victims of a terrorist act, if the terrorists' ends are just, then the act may be morally permissible.[19]

But while consequentialism may judge acts of terror permissible under some conditions, the likelihood of those conditions being met is vanishingly small.[20] And that vanishingly small likelihood is essentially reduced to zero when the terrorists use weapons of mass destruction, whose range is unpredictable and impossible to control, and which may have unknown long-term effects on the environment.

Agents of Death

Terrorists who want to poison a small number of victims are likely to use ordinary household or industrial chemicals. One well-known murder manual instructs readers in how to tamper with pharmaceutical capsules, how to contaminate food products, and how to use simple household chemicals to kill guests at a party. The author suggests testing poisons on drunkards rather than on stray cats, which he claims are surprisingly immune to poisons lethal to humans.

Criminals have followed the instructions in such manuals. A poisoning manual's recommendation that pharmaceutical capsules be filled with cyanide and returned to their original containers predated the ten known murders committed by this method. Four children and an adult were allegedly injured by eye drops contaminated with carbolic acid after a manual recommended this technique. In 1985 a man became ill from drinking a bottled beverage contaminated with urine, a technique recommended by a revenge manual in 1981. Cyanide-laced teabags were discovered in a grocery store four years after murder manuals suggested this method. If conducted on a larger scale, product tampering of this type could wreak havoc on a nation's economy.

Besides household materials, terrorists might use commonly available industrial or agricultural chemicals, some of which are highly toxic. The leak of methyl isocyanate at Bhopal, India, in December 1984 illustrates their deadly potential. Methyl isocyanate, used in the production of insecticides, is lethal in high concentrations. Union Carbide, the owner of the pesticide plant in Bhopal, concluded that the explosion was caused by a disgruntled employee, who added water to a storage tank. The water caused an explosion which released the poisonous gas. An official Indian investigation concluded that nearly 4,000 people died from exposure to the lethal gas and approximately 11,000 were disabled. This incident shows that a single angry worker can be extremely dangerous: more people were killed in this chemical release than in any single terrorist attack to date.[21]

Or terrorists might try to use chemical or biological poisons that have been developed as warfare agents. Because these poisons are more toxic than industrial or household chemicals, a smaller quantity is required to produce casualties. Anthrax, a biological agent, can be extracted from the soil. Chlorine, used as a chemical weapon in World War I, is widely available in industry. Chemical and biological agents are usually spread as aerosols: victims die when they inhale the poison. Unlike tanks, guns, and bombs, these agents have no effect on a nation's infrastructure. Both can be disseminated silently (no explosion and no noise). Both are relatively easy to hide.

Chemical and biological agents are often lumped together conceptually, but there are several important distinctions between them. *Biological warfare agents* are "living organisms . . . or infective material derived from them, which are intended to cause disease or death in man, animals and plants, and which depend for their effects on their ability to multiply in the person, animal, or plant attacked." *Toxins* are poisonous substances, usually produced by living organisms. *Chemical warfare agents* are chemical substances, whether gaseous, liquid, or solid, which are used for hostile purposes to cause disease or death in humans, animals, or plants, and which depend on direct toxicity for their primary effect.[22]

Differences between them are often described in terms of five factors: *toxicity, speed of action, specificity, controllability,* and *residual effects.* Biological agents are more toxic than chemical agents, slower-acting, sometimes more persistent (in the case of spore-forming agents), more specific (in that they may affect only a single species, whereas chemical agents affect all organisms), and less controllable, and may have longer residual effects.

Biological Agents

- *Toxicity:* Tiny amounts of biological agents can kill. Lethal doses of toxins are measured in micrograms (1×10^{-6} grams); those of biological agents are generally measured in picograms (1×10^{-12} grams).[23]
- *Speed of action:* Live biological agents act slowly. The agent must multiply in the victim before the victim will exhibit symptoms, a process that can take from several hours to several weeks. The incubation period of anthrax, for example, is from one to six days. Treatment must begin before symptoms appear. Toxins act immediately.
- *Specificity:* Biological agents are specific: most affect either plants, humans, or other animals exclusively.
- *Controllability:* The effects of biological agents are highly dependent on difficult-to-control variables such as meteorological conditions and terrain. The spread of contagious agents is particularly hard to control or predict because victims may infect others. Some biological agents may be spread not only by human travelers but also by animals such as migratory birds.[24]
- *Residual effects:* Most biological agents will not survive long in the atmosphere. Anthrax in spore form, however, persists in soil or structures for many years.

To be useful as warfare agents, biological organisms must possess the following characteristics: they must be highly infectious; they must be relatively resistant to the atmosphere; they must be sturdy enough to maintain viability and virulence during production, storage, transportation, and dissemination; and they must be capable of being produced in large quantities.[25]

Not all of these strictures apply to terrorist uses, however. If the terrorists' objective were to kill large numbers, these requirements would apply; but if the objective were to murder a small group, the terrorists might be satisfied with a toxin that could be injected rather than disseminated more widely. Terrorists might also opt for infecting victims with common food-borne contaminants like staphylococcus, salmonella, or shigella, which would make victims ill but probably not kill them. Resourceful terrorists might even try to engineer new lethal organisms.

Biological organisms developed for use as warfare agents fall into five categories: viruses, bacteria, fungi, rickettsia, and toxins.[26] For more details on biological warfare agents, see Tables 1 and 2.

- *Viruses* are often thought of as living organisms, but they are actually a collection of genes wrapped in a coat of protein. Unlike bacteria, they are unable to reproduce outside the tissues of the host (which may be an animal or a plant). When a virus enters a cell, the cell cannot distinguish the virus's genetic code from its own, so it replicates the genetic material of the virus, thereby causing infection. Viruses that have been considered for use as warfare agents include encephalitis, psittacosis, yellow fever, and dengue fever. Most viruses cannot survive outside the host, but some (such as rhinoviruses that cause the common cold) do survive long enough to be highly contagious. Fortunately, the most lethal viruses, such as Ebola, tend to be less easily transmitted than influenza or rhinoviruses.
- *Bacteria* are genuine living organisms capable of surviving outside the host, in some cases indefinitely. Bacterial

diseases considered for use in warfare include anthrax, brucellosis, glanders, plague, and tularemia.

- *Fungi* considered as warfare agents include *Pyricularia oryzae,* an easy-to-grow fungus that causes rice blast, a very destructive disease of rice plants. (For other antiplant agents, see Table 2.)[27]
- *Rickettsia* are parasitic microorganisms such as *Coxiella burnetii,* which causes Q-fever, a highly infectious disease that rarely kills but might be used to incapacitate victims. Rickettsia are highly infectious and are generally hardy in the airborne state, making them of interest as a warfare agent.
- *Toxins* are nonliving chemicals produced by living organisms. They do not replicate, but are by definition highly poisonous. Botulinum toxin is the most toxic substance known. Others are ricin (obtained from the castor bean plant); saxitoxin (obtained from contaminated shellfish); and venoms (obtained from reptiles or insects).[28]

No case of large-scale biological warfare has been confirmed, although during World War II Japan experimented with biological weapons on prisoners of war and dropped ceramic bombs containing fleas infected with bubonic plague on remote villages in China. Biological weapons are alleged to have been used in war in several other cases as well.[29]

The Convention on the Prohibition of the Development, Production and Stockpiling of Bacteriological (Biological) and Toxin Weapons and on Their Destruction (known as the Biological Weapons Convention or BWC) bans the development, production, acquisition, stockpiling, transfer, and use of biological agents and toxins (except for peaceful purposes). The Convention bans both agents and weapons systems. The BWC opened for signature on April 10, 1972, and entered into force on March 26, 1975. The United States had unilaterally renounced biological warfare in 1969. As of July 1, 1998, 140 countries were parties (signed and ratified) to the BWC and 18 countries had signed but not ratified the treaty. The BWC contains no verification measures. Negotiations have been under way since 1995 to develop a compliance and transparency protocol.

Chemical Agents

- *Toxicity:* Lethal or incapacitating doses of chemical agents are measured in milligrams (1/1000 of a gram).[30]
- *Speed of action:* Chemical agents can kill within minutes.
- *Specificity:* Chemical agents affect all living things.
- *Controllability:* As with biological agents, the effects of dispersed chemical agents are highly dependent on difficult-to-control variables such as meteorological conditions and terrain.
- *Residual effects:* Volatile chemical agents, such as sarin or chlorine, dissipate within minutes. Less volatile agents, such as mustard or VX, persist for hours or days. Mustard reportedly persists even longer when contained underground.

To be useful as warfare agents, toxic chemicals must possess the following characteristics: they must be highly toxic per unit of weight; they must be relatively resistant to the atmosphere; and they must be capable of being produced in mass quantities. Depending on the operation, these requirements would not necessarily apply to terrorist acts. For example, if the terrorists' aim was to kill or injure small numbers of people, large quantities would not be required.

There are four categories of chemical warfare agents: choking and incapacitating agents (including chlorine and phosgene), blood agents (including hydrogen cyanide and cyanogen chloride), blister agents (including mustard and Lewisite), and nerve agents (including tabun, sarin, soman, and V agents, plus the new novichok agents developed in Russia). For more details on chemical warfare agents, see Table 3.

In addition, terrorists might attempt to disseminate toxic agricultural chemicals, including organochlorine insecticides, organophosphorus pesticides, herbicides, and carbamates; toxic industrial chemicals, such as hydrogen cyanide, carbonyl chloride, cyanogen chloride, and arsine (all of which have been used as warfare agents in the past); and heavy metals, including arsenic, mercury, cadmium, and lead.

Confirmed cases of chemical warfare include the use of chemical agents by both sides in World War I (1914–1918); by Italy in Ethiopia (1935–1936); by Japan in China (1937–1945); by Egypt in Yemen (1963–1967); and by Iraq in Iran and Kurdistan (1983–1988).

The Chemical Weapons Convention (CWC) bans the development, production, acquisition, stockpiling, transfer, and use of chemical weapons. Production of small amounts of chemical agents is allowed, provided the purpose is peaceful (for example in medicine). The CWC opened for signature on January 13, 1993, and entered into force on April 29, 1997. As of July 1, 1998, 112 countries were parties (signed and ratified) to the CWC and 56 had signed but not ratified. The CWC requires that all parties destroy their chemical weapons and chemical-weapons production facilities, even if the facilities or weapons are located on the territory of another country. It includes unprecedentedly intrusive inspection measures, including routine inspections at declared facilities where chemicals covered under the CWC are stored or processed and short-notice challenge inspections at suspect sites.

Radiological Agents

Isotopes are atoms of the same element having different numbers of neutrons in their nucleus. For example, uranium-235 atoms contain 92 protons and 143 neutrons, while uranium-238 atoms contain 92 protons and 146 neutrons. Isotopes of the same element usually have nearly identical chemical and physical properties, but may vary greatly in their nuclear properties.

Radioactive isotopes are unstable. They spontaneously decay into new elements. In the process they emit particles or energy (usually in the form of alpha or beta particles, gamma rays, or neutrons). Eventually, as a result of one or more stages of radioactive decay, a stable (nonradioactive) end product is formed. For example, uranium-238 (the most common uranium isotope) decays into lead-206, a nonradioactive, stable element. But there are thirteen intermediate steps, each of which results in an intermediate unstable isotope.

"Radioactive half-life" refers to the time it takes for half of a sample of a particular isotope to decay. The larger the number of nuclei in the sample, the more intense the radiation. But the half-life depends on the identity of the isotope, not the size of the sample. If we start with a kilogram of tritium, which has a half-life of 12.26 years, in 12.26 years half a kilogram of tritium will remain, and in 24.52 years a quarter of a kilogram will remain. Isotopes with short half-lives produce high levels of radioactivity (for a given sample size); those with long half-lives are less dangerous to handle. Terrorists would probably try to obtain isotopes with relatively short half-lives.

External radiation causes harm only if it penetrates the body. Gamma rays, protons, and neutrons readily penetrate the body. Beta particles are significantly less penetrating, but energetic beta particles can penetrate the skin and cause severe burns. Beta emitters deposited within the body create a serious internal radiation hazard. Alpha particles are not penetrating, but pose a serious threat to health if ingested or inhaled.[31]

Internal exposure occurs when radioactive substances are ingested, inhaled, or absorbed through the skin. Plutonium is highly toxic, but only if ingested or inhaled. Weapons-grade plutonium (over 93 percent Pu-239) is an alpha emitter, meaning that little or none of its radiation is absorbed through the skin. Plutonium is more dangerous when airborne than in water or food because it is more readily taken into the body via the lungs than via the gastrointestinal tract. Only very small specks of it (between one and five microns in diameter) can be inhaled. Particles of this size do not form spontaneously, and more of them are formed during an explosion than during a fire.[32] Beta emitters, such as thorium-234 and bismuth-210, present a hazard to the skin. Gamma emitters are significantly more dangerous as external sources; they do not have to be ground up and disseminated, but they act only over short distances. Gamma emitters include strontium-90, cobalt-60, cesium-137, and most medical and industrial isotopes.

The unit of measurement for the radiation energy absorbed by the body is the *rad.* A rad is absorbed when an *erg* (a unit of energy) penetrates a gram of tissue. One *rem* is the damage to tissue caused by the absorption of a rad of gamma radiation. The more rapidly the radiation is absorbed, the greater the damage to tissue; the unit for measuring this is the *rem/hour.* "Acute exposure" refers to the radiation absorbed within twenty-four hours. For the effects on humans of various levels of acute exposure, see Table 4.

The U.S. government considered developing radiological weapons during World War II, but abandoned the project as impractical. Unlike chemical and biological agents, radioactive poisons act slowly. Depending on the source and concentration of ionizing radiation, the victim may feel no ill effects for months or even years. Moreover, radiological agents are difficult to disseminate in concentrations sufficient to cause death, radiation sickness, or cancer.

Comparing radiation dispersal devices with chemical weapons in the 1940s, U.S. government scientists noted that chemical agents can be stored in large quantities for long periods, are easier to transport than radiological weapons, and are less expensive. They were also available at that time in significantly larger quantities. Sources of gamma radiation cannot be stored in large quantities, their effectiveness decreases with time, and they require adequate shielding and cooling. And in the 1940s they were "extremely expensive" and available only in small quantities.[33] These same factors might lead terrorists to choose chemical weapons over radiological ones, but only if their objective was to kill large numbers of people. Dispersing gamma radiation (for example, by attacking a power reactor) could impose severe economic and psychological costs on a target population.

Nuclear Weapons

Nuclear energy is released when heavy nuclei are split into smaller nuclei in a process called *fission.* Energy can also be released when light nuclei are combined to form a heavier nucleus in a process called *fusion.* But fusion requires temperatures comparable to those at the core of the sun and extremely high pressures, so terrorist-produced nuclear weapons are unlikely to involve fusion.[34]

A chain reaction occurs when some of the neutrons released by fission strike other nuclei, causing them to undergo fission in turn. Some of the neutrons are scattered or captured (for example in nonfission reactions between neutrons and nuclei), but some cause additional fissions. A chain reaction is self-sustaining (or stationary) if there are enough neutrons to cause continuing fissions, and if the gain in neutrons through fission is exactly balanced by the losses to scattering and capture. This is the kind of chain reaction created in power reactors. The amount of energy released depends on the number of fissions per second.

Fissionable materials are isotopes whose nuclei can fission (split into two lighter nuclei, usually emitting neutrons and large amounts of energy) by absorbing fast neutrons, but cannot sustain a chain reaction. The uranium isotope U-238 (found in abundance in natural uranium) is fissionable. The term is often used incorrectly as a synonym for fissile material. *Fissile materials* are isotopes that can fission by absorbing a slow neutron (a neutron that has been slowed by collisions with nuclei) and can support a neutron chain reaction. Examples include U-235 and Pu-239.

In a multiplying (or divergent) chain reaction, the number of free neutrons gained through fission exceeds those which are lost. If sustained long enough, a multiplying chain reaction produces a nuclear explosion. The minimum quantity of fissile material capable of sustaining a chain reaction (once it has been initiated by a neutron stream) is called a critical mass. The magnitude of the critical mass depends on the type of fissile material, the shape of the fissile mass, the density of the material, and the pressure of substances that either absorb neutrons or affect their speed. In a subcritical mass fissions can occur but gradually die out because there are not enough neutrons to maintain the reaction. In a supercritical system there is enough fissile material to support a multiplying chain reaction and a nuclear explosion.

Uranium-235 and plutonium-239 are the most efficient fissile materials for use in nuclear weapons. U-235 occurs in natural uranium deposits, but makes up only about 0.7 percent of natural uranium. To sustain a multiplying chain reaction, the U-235 content must be enriched to at least 10 percent, an expensive and difficult process. At that level of enrichment, 4,000 kilograms of U-235 would be required to make a bomb; thus material enriched to 10 percent is impractical for bomb-making purposes. Uranium that has been enriched to more than 20 percent U-235 is called highly enriched uranium (HEU). U.S. weapons employ HEU enriched to greater than 90 percent U-235, which is called weapons-grade uranium; 25 kilograms of this is enough to make a bomb.[35]

Plutonium does not occur in nature at all, but must be produced in a nuclear reactor. Only 8 kilograms are needed to make a bomb; thus plutonium can be used to make smaller, lighter weapons. Unlike the case with uranium, all plutonium isotopes are fissile. But Pu-240 is less desirable in weapons than Pu-239; a bomb will have a higher assured yield if it has a high Pu-239 content.

In a fission weapon, the mass of nuclear material must be subcritical before detonation and supercritical after detonation. One of two techniques is employed: a gun-assembly device or an implosion weapon.[36] When a gun-assembly device is used, two or more subcritical masses are fired together rapidly to create a supercritical mass. Gun-assembly devices are conceptually relatively simple; if terrorists were to make a nuclear device, it would probably be a gun-assembly device. The drawback from the terrorists' point of view is that these weapons can only be made with HEU, which may be harder to obtain than plutonium.

In an implosion weapon, chemical high explosive is packed around a core of fissile material. The high explosive is then detonated uniformly to compress the subcritical mass into a dense supercritical mass. Implosion weapons can be made with either uranium or plutonium, but they are significantly more complicated to produce than gun-assembly devices. They require great precision to initiate the implosion in all segments simultaneously. While this type of weapon would be harder to make than a gun-assembly device, it cannot be entirely ruled out as a possibility for terrorists (see Chapter 4).

The purpose of the Nuclear Nonproliferation Treaty (NPT) is to prevent the further spread of nuclear weapons; to provide assurance, through international safeguards, that the peaceful nuclear activities of states that have not already developed nuclear weapons will not be diverted to making such weapons; and to promote the peaceful uses of nuclear energy. The NPT opened for signature on July 1, 1968, and entered into force on March 5, 1970. As of July 1, 1998, 186 countries were parties to the treaty. Parties that do not have nuclear weapons are obliged to negotiate agreements with the International Atomic Energy Agency (IAEA) for safeguards, provisions designed to detect and deter the diversion of nuclear materials from peaceful to weapons use. Each party negotiates its own safeguards agreement with the IAEA based on a model agreement. Parties that do have nuclear weapons (the United States, the United Kingdom, France, the Soviet Union, and China are the only countries allowed to possess nuclear weapons under the NPT) are obliged to reduce their nuclear arsenals, with the ultimate objective of general and complete disarmament. Iran and Iraq have both signed and ratified the treaty. India, Israel and Pakistan have not. India has strongly objected to the discriminatory nature of the NPT, in that there are different rules for the "haves" and "have nots."

Terrorism, as defined here, is an act or threat of violence against noncombatants, with the objective of intimidating or otherwise influencing an audience or audiences. Nuclear, chemical, and biological agents are ideal weapons of terror because they are inherently intimidating and mysterious. The outcome of their use—in terms of lives lost—is impossible to predict with certainty. What we can predict is that the radius of psychological damage would exceed that of injury and death.

Under very narrow circumstances, some moral theories permit the deliberate killing of innocents. But the same factors that make nuclear, chemical, and biological weapons unusually dreadful also make them morally special. In most possible uses of such weapons, terrorists would not know in advance how many victims would succumb. Nor would they know the long-term residual effects on the health of survivors, or on the environment.

From a military point of view, useful attributes of biological and chemical agents are that they penetrate structures, cover large areas, and do not affect infrastructure. Nuclear weapons are of interest because of their immense destructive power. What General Marshall called the "shock value" of the atomic bomb applies equally to chemical and biological agents. To military planners, shock is of secondary importance. To terrorists, however, shock is the central consideration. Dread is the terrorist's instrument for achieving influence.

Trojan Horses of the Body

The [atomic] bomb would be dropped from a height that would minimize radio-active poisoning in order to avoid any contention that poison gases were being used.

—Harvey Bundy, 1946

Nuclear, chemical, and biological weapons are considered barbarous even though, in the words of Louis Renault, it is difficult to distinguish the barbarity of any particular weapon from the cruelty inherent in all acts of war.[1] Soldiers wounded by ordinary explosives are left lying on battlefields with dangling limbs, pierced body parts, and deep psychic wounds. Why are nuclear, chemical, and biological weapons singled out as unusually cruel?

I believe a visceral fear of poisons plays a role, not only in the perception that chemical and biological agents are unusually vicious, but also in the belief that nuclear weapons are morally unacceptable. Conventional fire-bombing raids killed many more noncombatants in World War II than did the atomic bombs dropped on Hiroshima and Nagasaki, but it is the nuclear weapons that have inspired scholarly debate, and for many, contrition. There is little evidence that military leaders viewed atomic bombs as morally distinct from conventional ones before they were used.[2] But later General Leslie Groves, the military director of the Manhattan Project, was distressed to discover that large numbers of Japanese victims suffered radiation poisoning. Groves had been under the impression that blast and heat would be the only significant lethal effects of a high-altitude burst: that there would be minimal damaging effects from radioactive materials on the ground. He was disturbed, in other words, by the poisonous (radioactive) aspect of these weapons.[3] When the United States proposed deploying neutron bombs (which have minimal blast and thermal effects but enhanced radiation) in Europe in 1977, public outcry was so strong that President Carter had to relinquish the plan. Similar concerns about radiation forced the U.S. government to relinquish above-ground nuclear tests. In both cases it was the poisonous aspects of nuclear weapons that upset the public. Such weapons are especially terrifying because they are, in Michael Mandelbaum's words, "Trojan horses of the human body: they slip inside and then attack."[4]

Poisons have always been seen as unacceptably cruel. Livy called poisonings of enemies "secret crimes." Cicero referred to poisoning as "an atrocity."[5] But why do poisons evoke such dread? This question has long puzzled political scientists and historians.[6] One answer is that people's perceptions of risk often do not match reality: that what we dread most is often not what actually threatens us most.

Risk versus Dread

When you got up this morning, you were exposed to serious risks at nearly every stage of your progression from bed to the office. Even lying in bed exposed you to serious hazards: 1 in 400 Americans is injured each year while doing nothing but lying in bed or sitting in a chair—because the headboard collapses, the frame gives way, or another such failure occurs. Your risk of suffering a lethal accident in your bathtub or shower was one in a million.[7]

Your breakfast increased your risk of cancer, heart attack, obesity, or malnutrition, depending on what you ate. Although both margarine and butter appear to contribute to heart disease, a new theory suggests that low-fat diets make you fat. If you breakfasted on grains (even organic ones), you exposed yourself to dangerous toxins: plants produce their own natural pesticides to fight off fungi and herbivores, and many of these are more harmful than synthetic pesticide residues. Your cereal with milk may have been contaminated by mold toxins, including the deadly aflatoxin found in peanuts, corn, and milk. And your eggs may have contained benzene, another known carcinogen. Your cup of coffee included twenty-six compounds known to be mutagenic: if coffee were synthesized in the laboratory, the FDA would probably ban it as a cancer-causing substance.

If you smoked a cigarette, statisticians believe, you reduced your life span by five minutes. You took an especially grave risk driving to work: 110 Americans per day die in car accidents. If you drive a small car your risk of death is twice as high as if you drive a gas guzzler—but the gas guzzler contributes to greenhouse gases and global warming, and also increases the risk to others should you suffer an accident.

At work you are exposed to additional risks. Your risk of dying this year from a job-related illness or injury is 1 in 1,100 if you are a pilot; 1 in 2,300 if you work in mining or agriculture; 1 in 11,000 if you work for the government; and 1 in 23,000 if you work in manufacturing. But if you are now considering quitting your job to avoid danger, think again. Because of the increased risk of stress-related disease—including suicide and alcoholism—being unemployed is as risky as smoking ten packs of cigarettes a day.

Most people are more worried about the risks of nuclear power plants than the risks of driving to work, and more alarmed by the prospect of terrorists with chemical weapons than by swimming in a pool. Experts tend to focus on probabilities and outcomes, but public perception of risk seems to depend on other variables: there is little correlation between objective risk and public dread. Examining possible reasons for this discrepancy will help us understand why the thought of terrorists with access to nuclear, chemical, and biological weapons fills us with dread.

Prospect Theory

Decision theory assumes that people make decisions by weighting the utility of each possible outcome by the probability of that outcome to get its expected utility, and then choosing the course of action that will yield the highest expected utility. But people do not respond to risky choices in this way. Psychologists have shown that we overestimate the likelihood of rare events, such as shark attacks, and underestimate the likelihood of more common ones, such as heart attacks. This helps explain why people are enthusiastic about high-prize lotteries, although the prospect of winning is low. It also explains why people are willing to overpay for airline flight insurance purchased immediately before a flight. We also tend not to distinguish adequately between large numbers (for example, a hundred versus a thousand). This collection of observations is known as prospect theory.[8]

All these observations are relevant to understanding our fear of nuclear, chemical, and biological terrorism. While terrorism is still relatively rare, especially in the United States, it looms large in the public consciousness. We may be in danger of overvaluing and overreacting to minor incidents because of our tendency not to distinguish adequately between 100 deaths and 10,000. This last point is especially likely to apply if terrorists employ nuclear, chemical, or biological agents, because our perception of the loss will be augmented by the dread these weapons evoke.

Rules of Thumb

Most of us rely on rules of thumb in calculating risks. Rather than carefully weighing pros and cons, we use heuristic devices. Justice Stephen Breyer of the Supreme Court explains: "We simplify radically; we reason with the help of a few readily understandable examples; we categorize (events and other people) in simple ways that tend to create binary choices—yes/no, friend/foe, eat/abstain, safe/dangerous, act/don't act—and may reflect deeply rooted aversions, such as fear of poisons."[9] We feel a gut-level fear of nuclear, biological, and chemical terrorism, and we want to eradicate the risk entirely, with little regard to costs.

Michael Mandelbaum argues that opposition to chemical weapons results from a genetic aversion rooted in human chromosomes. Richard Price and Nina Tannenwald call this argument "implausible" because it is "premised on the assumption that there must be a rational reason for the taboo that can be deduced from the essential features of these weapons."[10] But a genetic aversion implies irrational instinct, not rational thought. Whether or not Mandelbaum is correct about the origin of the human aversion to poison, his point that there is such an aversion (whether genetically or culturally based) appears to be sound. Anthropologists have observed a cross-cultural aversion to poison and impurity.[11]

Catastrophic risks are disproportionately feared.[12] The media tend to focus on spectacular events, such as tornadoes, fires, drownings, murders, and accidents. A study of two newspapers on opposite sides of the United States found three times as many articles on homicides as on diseases, although diseases take about 100 times as many lives, and perception of risk was highly correlated with level of news coverage. And people tend to fear terrorism more than ordinary crime. Because Belfast is thought of as a terrorist city, it is considered more dangerous than Washington, D.C., although there are far more murders per capita in Washington than in Belfast.[13]

People tend to exaggerate the likelihood of events that are easy to imagine or recall.[14] Disasters and catastrophes stay disproportionately rooted in the public consciousness, and evoke disproportionate fear. A picture of a mushroom cloud probably stays long in viewers' consciousness as an image of fear.

People tend to ignore hazards that seem routine, such as indoor air pollution, but fear those which attract media attention, such as hazardous waste sites, which actually pose lower aggregate risks to human health. Terrorist incidents are high-profile events: they tend to be dramatic and receive attention from the media. All of these factors contribute to our fear of terrorism, and to the risk of public panic if terrorists should employ weapons of mass destruction (WMD).

Dread, Familiarity, and Exposure

Risks that are disproportionately feared tend to display a certain cluster of characteristics. Dangers that arouse the deepest dread include those which are uncontrollable and inequitable, may lead to global catastrophe, are likely to affect future generations, and for which exposure is involuntary. Disproportionate dread is also evoked by risks with delayed or invisible effects, and by risks that are new or unexplained by science. WMD terrorism would be expected to evoke maximum dread, since nuclear, chemical, and biological weapons have all these characteristics. Victims of WMD terrorism might not be certain they had been attacked. Those injured by the blasts of nuclear weapons would know it immediately, but people exposed to radiation (the poisonous aspect of nuclear weapons) or harmful chemicals might be uncertain at first.[15] The author of a declassified top-secret study of the effects of nuclear weapons predicted that, in the event of an atomic attack on the United States, "many people will . . . be extremely apprehensive about the possibility that they may have been exposed to lethal amounts of radiation. For many hours and perhaps days, people in and around the disaster area might fear that their lives are endangered by lingering radiation products or by other invisible toxic agents."[16] Similarly, epidemiologists might have a hard time differentiating a biological attack from a natural outbreak of disease.

The danger of radiation from nuclear power plants is an example of a high-dread risk with a potentially high level of exposure. The health effects of such radiation are poorly understood (even by specialists), exposure is involuntary, the consequences are delayed, and the effects are initially invisible. In one survey of perceptions of risk, experts ranked nuclear power number twenty on a list of thirty technologies and activities, while most non-experts ranked nuclear power as the most dangerous item on the list. Swimming is an example of a low-dread, known risk: exposure is voluntary, the swimmers have the illusion of control, and if accidents occur their results are generally visible and immediate. While the experts ranked swimming as a relatively high-risk activity, college students ranked it number thirty—the least risky of all.[17]

Disgust and Horror

In *The Anatomy of Disgust,* William Miller says that horror is "fear-imbued" disgust; it is a subset of disgust for which "no distancing or evasive strategies exist that are not themselves utterly contaminating." "What makes horror so horrifying is that unlike fear, which presents a viable strategy (run!), horror denies flight as an option . . . Because the threatening thing is disgusting, one does not want to strike it, touch it, or grapple with it. Because it is frequently something that has already gotten inside of you or takes you over and possesses you, there is often no distinct other to fight anyway."[18]

Radiation and poisons get inside us and inhabit us. We cannot physically remove them as we can a sword or a bullet; we cannot escape being defiled. In the aftermath of a conventional bombing campaign, we can run from collapsing structures, and we know immediately whether we have escaped. When poisons spread, however, we may not know whether we've been poisoned and we may not be able to escape no matter how fast we run.

Unlike conventional weapons, poisons—radiological, chemical, or biological—are ingested. They contaminate the air we breathe, the water we drink, the food we eat. Ingestion and nourishment, normally linked to life, become linked to defilement and death. Perhaps this is why Alberico Gentili argued in 1612 that poisoning is a "violation of nature," and why Hamlet calls the poisoning of his father "strange and unnatural."[19]

When a person ingests radioactive substances, the body falls apart: skin sloughs off, hair falls out. (Radiation poisoning is only one of the horrifying effects of nuclear weapons, but an important one psychologically.) Mustard blisters the skin and eyes and sears the lungs. The effects of nerve gas and biological agents are even more horrifying.

Evil and disgust go hand in hand. What is evil disgusts us; what disgusts us may seem tainted with evil. We often feel dread and indignation in the presence of evil. We feel disgusted by what is cruel. But indignation, Miller posits, is inadequate in the face of evil.

Like morality, much of what disgusts people is both culturally and temporally based. But some behaviors evoke disgust across cultures and time. Incest and arbitrary murder are arguably in the category of behaviors that are universally vilified. Disgust plays a role in maintaining those norms, Miller argues.[20] Poisons have always been seen as immoral, and as particularly unworthy of civilized peoples. Poisoning is "worthy of brigands and not of princes," Gentili tells us. John of Salisbury declared, "I do not find the right to use poison by any law at any time, although it is sometimes employed by the infidels." Poisoning is "condemned by the law of nature and consent of all civilized nations," Monsieur de Vattel wrote in 1758.[21]

Poisons and Paranoia

Women who commit murder are supposedly prone to use poison. "There seems to be almost unanimous agreement among criminologists that the woman who kills uses poison more often than any other means," Otto Pollak claimed in 1950 in a well-received book on the criminality of women. Examples from literature include Medea, Circe, Deinira, Lamia, and Arachne. Dominik Wujastyk reports that the "venomous virgin" or "poison damsel" is a stock character of Sanskrit drama and narrative. In the *Tome on Medicine,* a seventh-century Indian manuscript, the venomous virgin is described as "a girl who has been exposed to poison from birth, and who has thus been made poisonous herself. She kills a lover just by her touch or her breath. Flowers and blossoms wilt when they come into contact with her head. The bugs in her bed, the lice in her clothes, and anyone who washes in the same water as her, all die. With this in mind, you should keep as far away from her as possible." A twelfth-century Indian pundit warns: "If she touches you, her sweat can kill. If you make love to her, your penis drops off like a ripe fruit from its stalk."[22]

"The image of the woman who uses poison or is venomous is, above all, an image of female power and male fear of that power," Margaret Hallissy argues. "The envenomed or serpentine woman in particular becomes a strong metaphor for the woman who is too unusual for the ordinary man to handle."[23] Eve was the original poison woman, against whom Adam had no defense. Although she herself was a sly temptress, Eve succumbed to the serpent's wiles and to her own irrepressible desire for knowledge, for which humanity will forever be punished. These images, repeated in literature since ancient times, affect us subliminally, giving poisonous weapons an aura of feminine guile and of evil.

Poison is seen as the weapon of the weak, of those who are forced to rely on deception. It is unmanly and unvalorous, worthy only of infidels or women. It is a "guileful deed when we desire to make use of poison." Those who use poison have "no valor; all their trust is in drugs." To poison is to resort to "wicked wiles," "evil frauds," and "treachery," to engage in behavior that is "contrary to honor and righteousness"—and therefore unworthy of men.[24]

Poison, the weapon of the weak, the wily, and the unworthy, is distinguished from force, the weapon of the strong, the honorable, and the righteous. Only barbarians take more pleasure in these "evil and treacherous wiles than they do in arms," Gentili admonished. "The idea of the deceptiveness of women is essential to understanding the image of poison," Hallissy explains. "Poison can never be used as an honorable weapon in a fair duel between worthy opponents, as the sword or gun, male weapons, can. A man who uses such a secret weapon is beneath contempt."[25]

Women are by nature more deceitful than men, Pollak argued, in part because their anatomy allows them to feign or hide sexual feelings: "Almost all criminals want to remain undetected, but it seems that women offenders are much better equipped [physically] for achieving this goal than men." Fear of venomous women and their anatomy can extend to fear of women's ability to secrete poisons inside their bodies. Larry Wayne Harris, who was arrested (and subsequently released) for carrying out experiments on anthrax vaccine in 1998, is convinced that Iraqi women are smuggling ampoules of anthrax and bubonic plague into the United States inside their "body cavities." This is not dangerous for the women, he claims, because they have been vaccinated.[26]

Teresa Brennan argues that since the 1950s we have been living in a paranoid age. Horror of poisons, at least in its current manifestation, may partly reflect that paranoia. We are constantly polluting our world, she says, and feel a collective sense of guilt. Fear of being poisoned is a projection of that feeling of guilt. We fear being poisoned—by radiation, by pollution, by disease. Our fear may even be justified. But the fear is nonetheless a psychological projection.[27]

Horror of Chemicals

Chemical agents produce horrifying results. Nerve agents such as VX, for example, inhibit the enzyme acetylcholinesterase, which prevents nerves from firing. Cholinesterase-inhibiting agents cause the nerves to fire continuously. Within seconds or minutes the victims feel pain in their eyes, their vision may blur, and they may feel tightness in the chest. Secretions pour from the nose and mouth, and the victims begin gasping desperately for air. They may suffer vomiting or diarrhea or both at once. They begin to twitch and then to convulse uncontrollably. They may lose consciousness. If not treated immediately, they will soon stop breathing. A single drop of VX on the skin is sufficient to kill a person.[28]

The effectiveness of chemical agents in war depends largely on fear. Chemical weapons (CW) caused serious panic and disorganization among troops during World War I, although the actual mortality rate was low.[29] Of the casualties inflicted by gas, some 2 percent resulted in death; of those inflicted by conventional weapons, 25 percent resulted in death.[30] A similar mortality ratio was observed in the Iran-Iraq war. Iraq's attacks against civilians, however, were devastatingly effective. Chemical attacks on the city of Halabja (a Northern Iraqi town then held by Iran) in March 1988 killed between 3,000 and 5,000 civilians. Despite the relatively low mortality rate for CW used against troops, Iraqi officials reportedly attributed their victory to Iranian soldiers' terror of CW.[31]

Americans have lost faith in what experts say about public exposure to chemical hazards. Part of the problem is that experts often disagree about the effects of toxic chemicals on human health. Even the best scientific methods of assessing risk invariably depend in part on intuition and inference. And often the effects of chemicals on human health must be estimated indirectly, on the basis of data about other animals. "Controversies over chemical risks may be fueled as much by limitations of the science of risk assessment and disagreements among experts as by public misconceptions," the authors of one study conclude.[32] While doctors have studied the effects of chemical agents for decades, continued lack of understanding of low-level exposure has resulted in a public relations fiasco for the Pentagon concerning the "Gulf War syndrome," a mysterious set of ailments believed by some specialists to be caused by exposure to a mixture of chemical agents, other pollutants, and possibly untested vaccines.

Horror of Disease

Plagues and disease can be even more frightening than toxic chemicals. Ebola, investigated as a possible weapon by at least one country (Russia) and also by the Aum Shinrikiyo cult, kills 90 percent of its victims within a week. Early signs of infection include flu-like fever and joint pain. Soon connective tissue starts to liquefy, and blood may seep through the softened skin if pressure is applied. The victims choke on sloughed-off pieces of tongue and throat that slide into their windpipes, and their eyeballs fill with blood, which leaks down their cheeks. As victims near death, they become convulsive. Infectious blood splashing around during convulsions is a risk to any unprotected caregivers. Medical researchers initially believed that Ebola's transmissibility was limited by its instability in air. But doctors were able to isolate active virus in syringes of victims' blood that had been lying on a desk in Africa for nearly a month. Although U.S. Army medical researchers believe that in nature the disease is transmitted only through infected bodily fluids, the deliberate creation of an infectious aerosol might lead to a deadly weapon.[33]

"Epidemic disease has killed orders of magnitude more people than all of mankind's efforts combined," according to a U.S. government biologist. Fear of disease is a reasonable response to the threat. But people tend to fear unusual diseases more than well-known killers. Malaria, an ancient disease, kills a million people a year. The Marburg virus, discovered in 1967, has killed only 10 people; Ebola has killed 800 since its discovery in 1976. Yet it is Ebola and Marburg that have inspired terrifying books and movies. We respond to the likelihood of death once the disease is contracted, rather than to the compound—and low—probability of both contracting the disease and succumbing to its effects.[34]

Plagues and epidemic diseases, long ago thought to be instruments of divine vengeance, still petrify observers as well as the afflicted. During the sixth-century plague of Justinian, people believed that an encounter with a phantom of human form was a sign that they would soon fall victim. Sometimes the phantom appeared to them in dreams, sometimes in their waking hours. "And after the plague had ceased, there was so much depravity and general licentiousness, that it seemed as though the diseases had left only the most wicked," Procopius recounted. During the Middle Ages no one knew what caused disease; and ignorance made epidemics even more fearsome. When the source of disease was unknown, all substances seemed potentially dangerous, creating the impression of a malign force in the universe.[35]

Disease, with the fear it evokes, is a powerful force. It has repeatedly changed the outcome of war. The Peloponnesians left Attica not because they feared the Athenians, who were shut up in their cities, but because they feared the Plague of Athens, described by Thucydides in the second book of the *Peloponnesian Wars.* Hans Zinsser, writing in the 1930s, attributed the Roman victory over the Carthaginians in the Punic Wars to an epidemic, probably smallpox, that afflicted the Carthaginian army in 396 BCE. And a plague forced the Huns to give up their otherwise unimpeded advance toward Constantinople in 425 CE.[36]

A particularly frightening aspect of warfare or terrorism using biological weapons is that it may be hard to distinguish from a natural outbreak of disease. Although epidemiologists can, with difficulty, tell the difference, there is no worldwide database to facilitate analysis. Over the centuries people have sometimes falsely attributed the effects of biological weapons to natural outbreaks and sometimes blamed sabotage for naturally occurring epidemics.

The Black Death, which killed nearly a quarter of the population of Europe within fifteen years in the mid-fourteenth century, was believed at the time to be caused by supernatural forces. In fact, it was probably spread deliberately, as an instrument of human war. During a battle with the Genoese in 1346, Mongols hurled plague-infested cadavers over the walls into Caffa, a port city on the east coast of Crimea then controlled by the Genoese. The Genoese fled, but not before many were infected with plague. Their ship made frequent stops on its way to Genoa, spreading plague at each port. Eventually the disease spread across Europe.[37]

The most significant case of modern biological warfare was Japan's attacks against the Chinese during World War II. Japan experimented on human subjects, killing some 3,000 people at Unit 731, its main facility for biological warfare experiments. The victims either died during the experiments or were executed when they became too ill to be used as subjects any longer. An unknown number of American prisoners of war were among the human guinea pigs. Documents in Chinese archives describe biological attacks against 11 cities and some 700 victims of artificially disseminated plague. Some of the effects of Japan's biological attacks against China may have been erroneously attributed to natural causes.[38]

Many scientists believed Soviet claims that an outbreak of anthrax at Sverdlovsk in 1979, which killed 96 people, had been caused by tainted meat, not by an accident at a secret biological weapons facility. But in fact, as President Yeltsin later admitted, the outbreak had not been natural: pulmonary anthrax had been spread by an explosion at the plant, which was developing anthrax for weapons use in violation of the 1972 Biological Weapons Convention.[39]

Sometimes people suspect deliberate dissemination by governments when the most likely explanation is a natural outbreak. When hoof-and-mouth disease struck pigs in Taiwan in 1997 for the first time in eighty-three years, Taiwanese farmers suspected that China had deliberately introduced the disease to the island to damage its economy. Some Americans erroneously blame the CIA for introducing the human immunodeficiency virus (HIV), which has spread disproportionately among gay men and African Americans.[40]

Fear of Nuclear Weapons and Radiation

The U.S. officials involved in the decision to use atomic bombs against Japan in World War II deliberately took advantage of the horror such weapons inspire. Psychological factors were of "great importance" in the selection of targets for the bombs. The Target Committee concluded that the initial use of the bomb should be "sufficiently spectacular" that its "importance" would be recognized internationally; targets were to be chosen to obtain "the greatest psychological effect against Japan." According to a document prepared by the Joint Chiefs of Staff in 1946, atomic bombs might be used against industrial and population centers "with a view to forcing capitulation through terror and disintegration of national morale."[41]

The atomic bombs killed an estimated 110,000 civilians, significantly fewer than those killed in conventional firebombing during the war. And yet, among the living, the atomic bombs inspired significantly more moral doubt and dread. Part of this dread must have been due to the newness of the weapon. Part, undoubtedly, came from its sheer destructive power, the strange mushroom cloud, and the terrible fires that ensued.[42] But an additional factor was the invisible, mysterious poison the bombs produced. According to a declassified study for the U.S. government, "During the weeks following the atomic disasters, the distressing symptoms and sudden deaths from radiation sickness . . . produced strong emotional reactions among some of the survivors," worsening the "emotional disturbances evoked by the original disaster experience."[43]

After the war, fear of nuclear weapons and of fallout spread to the United States, starting a mania for building backyard bomb shelters. A 1951 study of the effects of nuclear weapons predicted that if "large-scale A-bomb attacks are launched against the United States, the psychological impact upon the American people might prove to be as shattering as the physical devastation." The Office of Civil Defense tried unsuccessfully to assuage fears by claiming that "protection [from fallout] is possible," and that "food and water would be available and usable."[44]

In 1983, much to sponsors' surprise, more than 100 million Americans watched "The Day After," a two-hour documentary dramatizing the effects of a nuclear attack against the United States. Worry about nuclear war reached a peak in late 1983 and 1984, then dropped significantly in 1985, and has continued to decline since 1987.[45]

Distrust of Government

Although worry about nuclear war has been subsiding, Americans fear radiation from nuclear power plants more today than in the past. In the 1950s nuclear power was considered benign. By the 1980s, however, it was one of the most dreaded risks.[46] One reason for this change is that people do not trust the government as much as they used to.

America was founded on the notion that government is not

entirely to be trusted, that its power must be held in check—even at the expense of efficiency. But there has been a marked decline in public trust in the government since the mid-1960s, and a corresponding loss in government legitimacy and power. There are many possible explanations.[47] Some blame the tendency for bureaucracies to expand and to promise more than they can deliver.[48] Others blame the information revolution, which, they say, has bypassed traditional hierarchies (family, business, media, government, military) and facilitated transnational enterprises (including gangsterism, factionalism, and terrorism).[49]

Another possible explanation is that government has in fact proved untrustworthy, at least when it comes to managing risky technologies. When the National Association of Science Writers surveyed American views of science in 1957, nearly 90 percent of those polled believed the world was "better off because of science"; 88 percent believed science was "the main reason for our rapid progress"; and 90 percent believed science had no negative consequences.[50] The public was enthusiastic about technological advances and economic expansion during the 1950s and 1960s, and early indications of links between these positive developments and damage to the environment were largely ignored.

In the 1970s this optimism about technology began to erode. A series of environmental disasters contributed to the public's loss of faith. In 1979 the Three Mile Island nuclear power plant in Pennsylvania experienced a partial meltdown, releasing a plume of radioactive material over hundreds of square miles. Although the level of radiation released was reportedly not dangerous, public opinion turned against nuclear power. In 1986 a more serious accident at a nuclear plant in Chernobyl, Ukraine, intensified the aversion. And nuclear power was not the only technological threat to the environment. In 1983 all 2,200 residents of Times Beach, Missouri, were evacuated because of dioxin poisoning. A decade earlier, waste oil contaminated with dioxin had been sprayed on the city to reduce dust. But it had killed pets and livestock and apparently increased birth defects and nervous disorders. The Environmental Protection Agency bought the entire town for $33 million. In a series of polls in the 1980s, in stark contrast to the 1957 poll, a quarter or more of those surveyed believed that technology would do more harm than good to the human race, or that its risks outweighed its benefits.[51]

The arms race also contributed to the loss of faith in government and science. During the Cold War both sides exposed their own citizens to health risks in a sometimes reckless pursuit of nuclear superiority. After decades of carelessness resulting in devastating chemical spills and radiation leaks, the Russian government has little credibility in the area of protecting its citizens' health. And recent revelations have raised serious questions about the past reliability of the U.S. government as well. We now know that in the 1950s the government repeatedly lied to Americans about the effects on health of open-air experiments, experiments that were carried out on civilians without their knowledge—let alone their permission. In one government experiment doctors were asked to inject unwitting subjects with substances known to be hazardous.[52]

Kristin Shrader-Frechette believes that scientists have contributed to the public's loss of faith in their work by presenting their opinions as established facts. She notes that when scientists "present their own educated (but controversial) guesses as science, they can jeopardize the credibility of science. The result can be the anti-science sentiment that is widespread today."[53]

This lack of trust in government and in scientists has significant implications for terrorism. Societies whose citizens feel alienated from the government and from one another are more vulnerable than more cohesive societies to terrorist violence and also provide more fertile ground for breeding of extremists. The United States and Russia both stand out in this regard. And if a terrorist attack should occur, especially one using nuclear, chemical, or biological agents, mistrust of governments and scientists would seriously complicate the task of controlling panic.

Poison weapons, Leonard Cole observes, have always been seen as "inherently sneaky, unfair, abhorrent," for reasons that are hard to determine. John Moon describes this mysterious revulsion against poison as deep, mysterious, and ultimately inescapable: It can be rationalized away, but cannot be exorcized.[54]

The technologies and activities we fear most are not necessarily the ones that are the most dangerous. Certain characteristics of risks tend to inspire fear that is out of proportion to the risks themselves. Nuclear, chemical, and biological agents possess all the characteristics that are conducive to disproportionate dread. They can lead to spectacular catastrophe. Their long-term effects on the human body and on the environment are unpredictable or unknown. They are mysterious, indiscriminate, uncontrollable, inequitable, and invisible.

Because they evoke such horror, these weapons would seem to be ideal for terrorists, who seek to inspire fear in targeted populations. The media's tendency to highlight terrorist incidents intensifies the fear, and citizens who do not trust their governments are unlikely to respond to government calls for calm. Panic is the likely result, and panic itself is dangerous.

Getting and Using the Weapons

Humans, regrettably, have used available technologies for destructive as well as for beneficial purposes throughout history.

—*Journal of the American Medical Association,* 1997

Nuclear, chemical, and biological agents, as we have seen, are inherently terrorizing. They evoke moral dread and visceral revulsion out of proportion to their lethality. The government of a country attacked with such weapons would have difficulty controlling panic. Because chemical and biological weapons are silent killers, an attack could occur at any time without warning. The first sign of a biological attack might be "hundreds or thousands of ill or dying patients," a U.S. government scientist warns.[1]

Despite the evident appeal of such weapons as instruments of terror, terrorists have seldom used them. Terrorists have never detonated a nuclear device. They have used chemical agents rarely—most often to poison foods—and biological and radiological agents more rarely still. Except for the chemical attacks carried out by the Aum Shinrikiyo cult in Japan in 1994 and 1995, there have been no cases of large-scale, open-air dissemination. What has held terrorists back? The answer involves both technical constraints, which I discuss in this chapter, and motivational or organizational constraints, which I will cover in Chapter 5. The technical hurdles would be considerable: acquiring the agent or weapon would present one set of difficulties; disseminating or exploding it would present another.

Chemical and Biological Agents

Terrorists might be able to acquire chemical or biological (CB) agents from governments favorable to their cause. CB agents are proliferating. In 1997 Secretary of Defense William Cohen estimated the number of countries with "mature chemical and biological weapons programs" at "about thirty," and the CIA claimed that around twenty nations had developed these weapons.[2] Iran, Iraq, Libya, North Korea, and Syria—all listed by the State Department as supporters of terrorism—are believed to possess chemical weapons and at least some biological weapons. Iraq's CB programs are quite extensive (see Chapter 7). The small quantities of CB agents required for an attack would make it very difficult to track the flow of the weapons or their component chemicals to terrorist groups.

Terrorists also might be able to steal CB agents from national stockpiles. In Albania in 1997, according to an Albanian military official, antigovernment bandits stole chemical weapons and radioactive materials from four army depots. The stolen materials, the official warned, posed serious health hazards.[3]

Russia's security for chemical weapons is particularly problematic. The storage sites for chemical weapons were revealed in the newspaper *Rossiyskaya Gazeta* in January 1994, and in 1995 the army chief of staff, General Kolesnikov, expressed concern that publication of the locations increased the risk of theft. Kolesnikov also warned that the increases in crime in Russia are worsening the risk of "chemical weapons attacks."[4]

Because chemical and biological agents are relatively easy to produce, a single person with the right expertise could design an entire weapons program. Thus terrorists might acquire CB weapons by taking advantage of the "brain drain"—the prospect that weapons scientists will sell their expertise to the highest bidder. "Because of the deteriorating condition of the military-industrial complex in the former Soviet Union, many specialists in the field of chemical weaponry do not have enough sources of income to support their families and are ready to go anywhere to earn money," Vil Mirzayanov, a Russian chemical weapons scientist, said in 1995. Russian physicists are reportedly providing consulting services to missile and nuclear energy programs in Iran and

Pakistan. One physicist reportedly claimed his conscience did not trouble him at all, since, with so many defense specialists now out of work, "If I had not agreed, they would have just found someone else." And Russia is by no means the only country where disaffected or underemployed weapons scientists can be found. Libya has reportedly tried to hire scientists formerly employed in developing biological agents for South Africa.[5]

Terrorist groups that include—or hire—trained chemists would have no difficulty producing chemical agents. A report by the U.S. Office of Technology Assessment claims that "the level of technological sophistication required . . . may be lower than was the case for some of the sophisticated bombs that have been used against civilian aircraft."[6] Many of the components of these agents are widely sold for industrial purposes. For example, thiodiglycol, an immediate precursor to mustard agent, is used to make ink for ballpoint pens. All that is required to produce mustard from this material is a simple acid that also is easy to obtain.

Pathogens that could be used as crude biological weapons—such as the common food poisons salmonella, shigella, and staphylococcus—are readily available at clinical microbiology laboratories. Terrorists could also produce more deadly agents. Knowledge of microbiology, and of its potential applications to weapons, is increasingly widespread. Kathleen Bailey, after interviewing professors, graduate students, and pharmaceutical manufacturers, concluded that several biologists with only $10,000 worth of equipment could produce a significant quantity of biological agent. The requisite equipment would fit in a small room, she claims, and "the glassware, centrifuges, growth media, etc., can all be manufactured by virtually any country."[7]

Detailed information about how to set up a chemistry laboratory and how to order chemicals without arousing suspicion is published in manuals on poisoning. One manual instructs readers how to disseminate chlorine (a poisonous gas widely used in industry, which was used as a weapon during World War I) in a crowd. In 1982, the year of the first widely reported incident of tampering with pharmaceuticals, the Tylenol case, only a few poisoning manuals were available, and they were relatively hard to find. Today, how-to manuals on producing chemical and biological agents are advertised in paramilitary journals sold in magazine shops all over the United States, as well as on the Internet.[8] Although criminals are known to have used such manuals in plotting crimes, publishers maintain that publication of murder manuals is protected by the First Amendment.[9]

One of these manuals, *Bacteriological Warfare: A Major Threat to North America,* is described on the Internet as a book for helping readers survive a biological weapons attack. But in fact it also describes the reproduction and growth of biological agents and includes a chapter on "bacteria likely to be used by the terrorist." The book is sold over the Internet for $28.50 and is reportedly advertised on right-wing radio shows. Its author is Larry Wayne Harris, the former member of neo-Nazi organizations who ordered three vials of the bacterium that causes bubonic plague.

Once terrorists managed to acquire chemical and biological agents, they would face the question of how to use them against their targeted populations. Chemical agents are relatively easy to disseminate in enclosed spaces, and they are significantly easier than biological agents to spread over large areas. Key differences for the purposes of dissemination are that chemical agents are volatile (that is, some of the agent will spontaneously form a poisonous gas), while biological agents are not; and most biological agents are susceptible to humidity, desiccation, oxidation, air pollution, heat and shock (such as from explosions), and ultraviolet light, while chemical agents are not. (Anthrax in spore form is less susceptible to these insults than other biological agents.)

The most likely way to spread chemical agents in air would be to disseminate them in enclosed areas. If a suitcase full of nerve agent were opened into the air intake ducts of a building, many of the people inside would probably die. Even less of the agent would be needed to poison passengers on an airplane. Dissemination in open areas would be considerably more difficult. For example, a specially equipped car might be used to spread chemical agent in city streets. Some terrorists might have access to planes or helicopters fitted with crop sprayers, or even (probably only for state-sponsored groups) bombs or missiles.

As for biological agents, the U.S. Army has conducted tests of their dissemination in populated areas using nonlethal microorganisms to simulate biological agents. Six of these tests were conducted in San Francisco in 1950. A ship used an aerosol spray device to release two simulant agents (*Bacillus globigii* and *Serratia marcescens*) at various distances from shore. Air samples were taken at forty-three locations around the Bay Area. The Army's analysts concluded that it was feasible to attack a seaport city by disseminating biological agents from a ship offshore, and that success or failure would largely depend on the meteorological conditions at the time of the attack.[10]

The Army also conducted a test on the New York City subway in June 1966, employing another simulant *(Bacillus subtilis variant niger).* Technicians dropped light bulbs filled with bacteria into the system, either through ventilating grates or onto the roadbeds as the trains entered or left the station. The bulbs broke, releasing the bacteria. Aerosol clouds were momentarily visible after the bulbs were broken, but were ignored by most passengers, some of whom merely brushed off their clothes. Army scientists concluded that a large portion of the working population in downtown New York City would be exposed to disease if pathogenic agents were disseminated in several subway lines during rush hour.[11] New York City authorities were not informed of the test until 1980, when the Army's findings were made available to the general public. The results of this test in particular are quite frightening, but this scenario would require drying and milling or coating of the agent, tasks that might be beyond the capability of terrorist groups.

The Army also tested anti-animal and anti-crop agents, presumably spread with a crop duster. Nonbiological simulants were used in the anti-animal tests, but live agents, including wheat rust and rice blast, were used against crops. The tests were carried out until 1968, principally in Minnesota and Florida.[12]

The three ways of disseminating CB agents tested by the Army—from ships near seaports, on the subway, and with crop dusters—illustrate the kinds of low-technology attacks some terrorists might like to carry out. But it is not clear that they would be able to do so. Agents that kill humans or livestock must be disseminated in the form of aerosols that are respirable: that is, that can be taken into the body through the lungs. To be respirable, particles must be between one and five microns in diameter. Experts disagree about whether terrorists would be able to create such aerosols. Some claim that any college-trained molecular biologist is capable of killing hundreds of thousands of people in a single attack. Others cite Iraq as evidence that even governments have trouble overcoming the technical difficulties of dissemination, implying that terrorists without state sponsors would certainly fail. The Aum Shinrikiyo attacks in Japan lend credence to the latter view, but the truth is probably somewhere between these two viewpoints.[13] Terrorists might not be able to maximize efficiency in designing and building their weapons, but if they used massive amounts of agent it would hardly matter that less than one percent of the agent was effective. It would hardly matter to the victims, at any rate.

Biological agents can be disseminated in liquid or powder form. While producing liquid agent is relatively easy, disseminating it as an infectious aerosol is not. Dry powders can be disseminated far more easily. High-quality powders are complicated to make, however, involving skilled personnel and sophisticated equipment. Milling these powders, according to a U.S. government scientist, "would require a level of sophistication possessed [only] by some state sponsors of terrorism"; this implies that terrorists without state sponsors would have a difficult time indeed.[14]

Terrorists without state sponsors would be unlikely to master the technology required for dispersing liquid agents over broad areas unless they were able to hire skilled scientists trained in government biological weapons programs. But if terrorists used large quantities of liquid agent, then even an off-the-shelf sprayer could pose a significant threat—not to an entire city, but to, for example, passengers on a train.

Besides releasing CB agents into the air, terrorists might try contaminating a water supply or tampering with pharmaceuticals or foods. If the aim was to damage the target country's economy or to injure a particular company, the terrorists would only have to poison a fraction of the target foods in order to create fear. In the United States, cola, milk, and baby food could be particularly vulnerable targets.

Water supplies require collection, treatment, storage, and distribution. The two common sources from which drinking water is collected are ground water (wells) and surface water (rivers and lakes). Wells are more commonly used in rural areas for individual homes, and surface water is more common in big cities. Well systems make easy targets, but they tend to be small. Most surface water systems in the United States are so large that any contamination would probably be diluted too quickly to be harmful. This may not be the case in other countries.

Treatment plants are not very vulnerable, according to an Army study, because an attack would be detected before the contaminated water was widely distributed. But a case of contaminated water in Washington, D.C., in 1994 (not caused by terrorists) suggests that this is not necessarily true. Authorities did not detect high levels of cryptosporidium growing in Washington's water supplies until after the water had been distributed. Residents were instructed to drink only bottled water until authorities could purify the water. A similar outbreak in Milwaukee killed some 100 people. Chemical agents can also be removed from water, although some agents require more elaborate procedures than others.[15]

From treatment plants, water is usually pumped to storage tanks or reservoirs. Attacking a reservoir has been shown to be both costly and ineffective, especially with large reservoirs, in which dilution would soon render contaminants inactive. In the United States water in large reservoirs is routinely treated and tested.[16] But remotely located small reservoirs could nonetheless be vulnerable. There is little security at some of these sites, and the water quality is monitored less often. Attacking distribution networks would require significantly lower quantities of contaminants than attacking an entire reservoir, but the number of people put at risk would decrease commensurately. Distribution networks might be vulnerable, however, to terrorists whose goal was to attack a particular symbolic target or a building of strategic importance.[17]

Radiological Weapons

Detonation of a nuclear device is the least likely form of terrorism involving weapons of mass destruction. But technical challenges would not prevent less dramatic uses of radiological materials, which, although unlikely to kill or injure many people, could impose heavy financial and psychological costs on the targeted government.

Radioactive isotopes can be found at a number of diverse facilities, including hospitals and industrial plants, and in waste from nuclear power plants. While industrial isotopes are found in higher quantities in industrialized countries, nuclear waste is found all over the world. Unlike chemical, biological, or nuclear weapons, radiological weapons (now generally called radiation-dispersal devices or RDDs) are not banned by any international treaty.

The U.S. Department of Energy maintains a database of cases of smuggling of nuclear materials reported in the press, including not only fissile materials but also nonfissile radioisotopes. Most thefts of nuclear materials have occurred in the former Soviet Union (see Chapter 6), and most have involved nonfissile radioisotopes. But other countries are by no means immune to loss or theft. The U.S. General Accounting Office estimates that, between 1955 and 1977, unaccounted-for special nuclear (that is, fissile) materials totaled in the thousands of kilograms.[18] These losses are almost certainly attributable to accounting errors, but they indicate a distressing lack of care.

Terrorists who obtained radiological materials would next face the technical hurdles of turning them into RDDs. In experiments conducted in the 1940s and 1950s, the U.S. military found that disseminating gamma-emitting radiological agents in air involved enormous difficulties because of the heat generated by the material and the problem of dissipation.[19] Gamma emitters require heavy shielding, and some have to be used immediately because of radioactive decay. The military contemplated dispersing the substances using artillery shells, mortar shells, aircraft, and "fission aerial bombs or fission projectiles." The effects would be highly dependent on weather conditions and terrain. Cities would be very costly to attack. Stafford Warren found that "something approaching 100 times greater concentration" would be required for built-up areas because structures would absorb a large portion of the radiation.[20]

The U.S. government concluded after years of study that RDDs were not militarily useful. Iraq, too, apparently renounced them as impractical. But radiological weapons might meet some terrorists' objectives, especially if the terrorists were more interested in imposing financial costs or creating panic than in killing. Even if the terrorists used a crude dissemination device or a quantity of radioactive material too small to present a serious threat to health, fear of radiation could cause panic. When the United States was considering producing RDDs in the 1940s and 1950s, the demoralizing of personnel was explicitly mentioned as one of the "specific uses" of radiological warfare.[21]

The costs imposed on a targeted facility or government would be immense. In 1966 an American B-52 bomber collided with another plane over Spain. The conventional explosives in two of the B-52's hydrogen bombs detonated and dispersed plutonium over several hundred acres. Workers had to plow up 285 acres and remove soil from 5.5 acres, which was shipped back to the United States. The cost of the cleanup, including finding and recovering the bombs, was $100 million (in 1966 dollars), including $10 million for recovery of a bomb that fell into the sea.[22]

A relatively easy way for terrorists to disperse radioisotopes (or chemical or biological agents) would be through the ventilation system of a building. Terrorists might also use sprayers or blowers to disseminate fine powder, or release the powder from the top of a tall building. The U.S. military also considered placing radioactive material in enemy water supplies, which might work for small-scale operations but would not be very effective where water is kept in large reservoirs.[23]

After police seized a series of small nuclear caches in Germany in the spring and summer of 1994, numerous stories in the press warned that terrorists could use stolen plutonium to poison water supplies. It is possible to make soluble plutonium compounds, but in fact metallic plutonium sinks. Moreover, plutonium is more readily absorbed by the lungs than by the gastrointestinal tract, so it is an inefficient agent for poisoning water or food.

Nuclear facilities such as power plants may be attractive targets for terrorists. People who live near these facilities are highly sensitized to the health risks of radiation. The regulatory procedures that have been devised to deal with reactor accidents are more rigorous than required for public safety but not rigorous enough to reassure the public. For example, if a radiation level five times higher than background were detected outside Rocky Flats in Colorado, officials would be required to evacuate the city even though scientists consider this level harmless.[24] A terrorist whose objective was to create panic and wreak havoc (rather than to kill people) could take advantage of these tight regulations: a relatively small amount of radioactive material disseminated near Rocky Flats would be enough to force the city to evacuate—with significant economic and psychological repercussions for residents.

Power reactors are also vulnerable to sabotage. Spent fuel rods contain cesium-137 and other gamma-emitting isotopes. Sabotage would probably require the complicity of plant employees, but terrorists might also be able to damage a reactor or a tank holding radioactive waste by placing explosives outside the fence or attacking the plant by plane.[25] An explosion would be significantly more dangerous than a fire. Antinuclear terrorists in Germany have already tried to sabotage the transport of spent fuel rods, although there is no evidence that they were trying to steal the rods.[26]

Nuclear Weapons

Only the most sophisticated terrorist groups would be likely to consider manufacturing their own nuclear weapons. For these groups, the binding constraint would probably be the acquisition of fissile materials, which are much more highly protected than nonfissile radioactive isotopes. A spate of thefts of nuclear materials in the former Soviet Union makes it clear that the current system for protecting fissile materials is inadequate.

The former Soviet Union is not the only potential source of fissile materials. In May 1997 security officers at Rocky Flats, a nuclear facility fifteen miles from Denver, Colorado, told federal investigators that security had grown so weak that terrorists would be able to penetrate the facility. Large amounts of weapons-grade material are stored at the plant, which produced plutonium during the Cold War. The plant's director of security and safeguards had quit "in disgust" a month earlier, claiming that he could no longer ensure the safety of Denver's citizens. He also warned that an antigovernment group called the Montana Militia had tried to recruit members from among the plant's guards. Although the recruitment attempts were reportedly unsuccessful, they suggest that antigovernment groups may be interested in nuclear or radiological terrorism. A government inquiry also revealed security lapses at other U.S. nuclear sites.[27]

In the unlikely, but not impossible, event that terrorists managed to buy or steal a sufficient quantity of fissile material, a key question is whether they could make a detonable nuclear device, with or without state sponsorship. A group of designers of nuclear weapons was commissioned in 1987 to consider this possibility. The group concluded that building a crude nuclear device was "within reach of terrorists having sufficient resources to recruit a team of three or four technically qualified specialists," with expertise in "several quite distinct areas [including] the physical, chemical and metallurgical properties of the various materials to be used . . . technology concerning high explosives . . . electric circuitry; and others."[28] Terrorists might even be able to detonate plutonium oxide powder without actually making a bomb, the designers said, although such an operation would be extremely dangerous and would require "tens of kilograms" of material. The new availability of nuclear material on the black market is therefore particularly troubling.

South Africa's secret nuclear program, closed down in 1989 and revealed to the world by President de Klerk in 1993, provides some insight into how a sophisticated terrorist group could build a nuclear bomb if it had access to fissile materials. The South African experience makes clear that "virtually anybody can make a bomb," according to Ambassador Thomas Graham: "The number of people required is relatively small, and a wealthy backer of a terrorist organization could provide the funds. We're talking about millions—not billions—of dollars, and hundreds—not thousands—of people (including support staff)."[29]

While enriching the uranium required a large infrastructure and technical expertise, South African officials demonstrated the ease of making nuclear weapons once the highly enriched uranium (HEU) is in hand. Most of the equipment required to make the bombs was easy to procure covertly. By the time de Klerk canceled the program, South Africa had acquired a secret plant for enriching uranium for bombs; a stockpile of weapons-grade uranium; and six gun-assembly fission weapons (the kind of weapon the United States used against Hiroshima).[30]

Waldo Stumpf, the head of South Africa's Nuclear Energy Corporation, estimated in 1994 that the entire project, including enriching the uranium, cost $200 million. "The nuclear deterrent programme was considered to be a far more cost-effective alternative to the development of, for example, a fighter aircraft capability," Stumpf said. It cost "less than 5 percent of the defense budget at the time."[31] Costs would have been significantly lower had South Africa been able to purchase the HEU. Thus, if a terrorist group had access to HEU, the cost of producing the bomb could be significantly less than $200 million.

Whenever possible, the South African technicians used simple machines not controlled for export. For example, they used a two-axis machine tool (designed for making two-dimensional shapes) to create the three-dimensional shape necessary for the gun-assembly device.[32]

In the early 1980s the project employed about 100 people, more than half of them in administrative support and security; 40 people worked in the weapons program, David Albright reports, but only 20 were actually building the weapons. By 1989, when the program ended, the workforce had risen to 300, half of whom worked in the weapons program.[33]

The number of people actually involved in making the weapons is shockingly small, but it is significantly more than the "three or four technically qualified specialists" envisioned by the group of nuclear weapons designers. Nonetheless, South Africa's experience shows that some terrorist groups (for example Hezbollah) have sufficient financial resources, although they apparently lack the requisite expertise and/or the requisite fissile material.

Instead of building their own weapon, terrorists (or a terrorist-supporting regime) might try to steal or buy a nuclear warhead. Stealing a warhead would require overcoming security at a site where weapons are stored or deployed, taking possession of the bomb, and bypassing any locks intended to prevent unauthorized detonation of the weapon.

This would clearly be easier if the terrorists were able to obtain the assistance of insiders (whether through persuasion, coercion, or bribery). Reports occasionally surface about former Soviet soldiers who are willing to sell warheads, and about terrorists' alleged interest in purchasing them, but no instance of an actual sale has been confirmed.[34]

Russia's approximately 6,000 long-range strategic weapons are protected by locks, making it impossible—at least in principle—to launch them without high-level authority. But thousands of shorter-range tactical weapons have less sophisticated protections or no locks at all. These short-range weapons would be both easier to steal and easier to detonate.

A leaked CIA report warns of the potential in Russia for "conspiracies within nuclear armed units." The report attributes the increased risk of conspiracy to deteriorating living conditions and morale, "even among elite nuclear submariners, nuclear warhead handlers, and the Strategic Rocket Forces." Submarines are a particular source of concern: Russian naval officers know a lot more about nuclear weapons than their ground-force counterparts, since they must be able to operate autonomously at sea.[35]

General Aleksandr Lebed, a former head of President Yeltsin's Security Council, has claimed that most of the atomic demolition munitions or "suitcase bombs" in the former Soviet arsenal are unaccounted for. It is not possible at this point to confirm or deny Lebed's claim, but these bombs, if they exist, would be perfect terrorist devices because they would be small enough to be carried by one man.[36]

The Case of Aum Shinrikiyo

The most successful and most publicized terrorist use of weapons of mass destruction to date was the case of the Japanese cult called Aum Shinrikiyo. In our attempt to understand modern terrorism it is worth taking a detailed look at this cult and its leader.[37]

Shoko Asahara, whose original name was Chizuo Matsumoto, was born on the southern Japanese island of Kyushu. His father wove straw floor mats (tatami) for a living, and was barely able to feed his family. The family of six lived in a small shack with a dirt floor. Asahara was blind in his left eye and only half-sighted in his right. Sighted children taunted him, and he in turn taunted those whose vision was worse than his own. At the boarding school for blind children that he attended, he tortured and tricked his classmates into performing services for him, and duped them in financial scams. From childhood on, Asahara was determined to make money.

After high school Asahara founded a clinic, where he charged high fees, the equivalent of up to $7,000, for a three-month course of treatment, which included yoga, "herbal tonics," and acupuncture. He also peddled his treatments to senior citizens in Tokyo's top hotels, charging them thousands of dollars to "cure" their rheumatism with tonics made of ingredients such as dried orange peel and alcohol. Eventually he was arrested for fraud, but the court-imposed fine of about $1,000 was a tiny fraction of his profits.

In 1984 Asahara formed a company called Aum Inc., which produced health drinks and ran yoga schools. Like many Japanese people in the 1980s, Asahara developed an interest in spirituality. He traveled to India, where he perfected his meditation technique, learning, he claimed, to levitate. He also claimed he could pass through solid walls, intuit people's thoughts, and chart people's past lives.

In 1986 a Japanese New Age journal, *Twilight Zone,* ran pictures of Asahara apparently suspended in midair in the lotus position. Attendance at his yoga schools rose, and with the profits Asahara was able to open schools throughout Japan. Soon afterward he claimed he heard a message from God while meditating, telling him that he had been chosen to lead God's army. At about the same time he met a radical historian who predicted that Armageddon would come by the year 2000. Only a small group would survive, the historian told Asahara, and the leader of that group would emerge in Japan. Asahara immediately cast himself as that leader. He began planning for Armageddon and his leadership role, and changed the name of his company to Aum Shinrikiyo (Aum Supreme Truth).

Asahara chose Shiva the Destroyer as the principal deity for Aum Supreme Truth, though he also emphasized the Judeo-Christian notion of Armageddon. At a seminar in 1987 he made his first prediction: nuclear war would break out between 1999 and 2003. There were fewer than fifteen years to prepare, he warned. Nuclear war could be averted, but only if Aum opened a branch in every country on earth. Those who were spiritually enlightened would survive even a nuclear holocaust.

Asahara found new ways to expand his revenues. He encouraged his students to cut off all ties with the outside world and to hand over all their assets as a way to foster their spiritual development. Asahara sold clippings from his beard for $375 a half-inch and his dirty bathwater (called Miracle Pond) for $800 a quart. He also opened a compound near Mount Fuji, which eventually included a very expensive (and hazardous) hospital, as well as laboratories for research on weapons of mass destruction.

Many of those attracted to Asahara's promise of spiritual enlightenment were scientists, physicians, and engineers from Japan's top schools. One was Seiichi Endo, who had been trained at Kyoto University and was working at the university's viral research center. Another was Hideo Murai, a brilliant astrophysicist who became Asahara's "engineer of the apocalypse." According to David Kaplan and Andrew Marshall, who have studied Aum Shinrikiyo, "The high-tech children of postindustrial Japan were fascinated by Aum's dramatic claims to supernatural power, its warnings of an apocalyptic future, its esoteric spiritualism."[38]

Asahara was also able to attract Russian scientists and engineers to the cause. Russia was like a supermarket for Asahara. He bought weapons and training, and he recruited among Russia's scientific elite, who were now unemployed or unpaid and seeking new missions. During the early 1990s Russia's Minister of Defense Grachev reportedly paved the way for three groups of Aum Shinrikiyo cult members to spend three days with military units in the Taman and Kantemirov divisions, where, for a fee, they were trained to use military equipment. The cult reportedly also received substantial assistance from Oleg Lobov, who was then the head of President Yeltsin's Security Council. The Russian Prosecutor General's office is investigating allegations by a Japanese cult member that Lobov provided assistance to the poison gas program.[39]

Asahara became obsessed with WMD, in part because he was

convinced that the CIA planned to use such weapons against Japan. Moreover, he believed his group would need every available weapon to survive Armageddon. He put his scientists to work on developing WMD. He also sent a deputy on several trips to Russia to purchase weapons and scientific assistance.

Soon these efforts began to pay off. Asahara directed a series of attacks using biological agents. Cult members have confessed to carrying out nine such attacks, including at the Japanese Diet, at the Imperial Palace, elsewhere in Tokyo, and at two American naval bases: Yokosuka and Yokohama. In April 1990 members reportedly drove through the city of Tokyo with a convoy of three trucks outfitted to spray botulinum toxin. The convoy drove by the American naval bases, where they attempted to spread the poison, and to Narita airport. No botulinum was reported detected, however, during any of these attacks, and no one was reported ill. Next the cult tried spreading anthrax. In July 1993 members tried to disseminate anthrax at several sites, including near the Imperial Palace and near the Diet. They again fitted a truck with a special sprayer and drove it around the city. These attacks also failed.[40]

The cult tried spreading anthrax from the roof of its headquarters building. Members put a steam generator on the roof and poured anthrax spores into it, then turned on a sprayer and fan and waited to see the results. Small birds died, and the generator emitted a putrid smell that permeated the neighborhood. It smelled like burning flesh, one resident told the press. When inspectors went to the Aum building to ask questions, they were told that the smell was from a mix of soybean oil and perfume—Chanel No. 5—which the cult was burning to purify the building.[41]

The group was more successful with chemical weapons. In June 1994 residents of Matsumoto, a mountain resort a hundred miles west of Tokyo, noticed a strange fog hugging the earth. Soon thereafter some residents experienced nausea, vomiting, pain in their eyes, and difficulty breathing. By the next morning 7 people had died and 200 were ill. A total of 600 eventually became ill. Dogs lay dead in the streets, and dead fish floated in a nearby pond. Doctors investigating the incident noted that victims had markedly reduced levels of acetylcholinesterase (an enzyme necessary for proper functioning of the nervous system), a sign that they had been exposed to toxic organophosphate-based insecticides or nerve agents. But even after traces of sarin, a nerve agent, were found in the pond, authorities did not suspect terrorism. No terrorist group had claimed responsibility, and the idea that terrorists could be responsible for such a heinous crime seemed too far-fetched to be believed. In fact, as subsequently became known, Aum Shinrikiyo had carried out the attack. Asahara's intention was to poison three judges living in the area. The judges survived the attack when the wind changed direction, but the poison spread over the town.[42]

Nine months later, on March 20, 1995, the cult used sarin again, this time in a deadly incident that attracted media attention. By this time the police were closing in on the group, and its scientists had to work fast. The terrorists placed hastily made sarin-filled polyethylene pouches on five Tokyo subway cars, and then punctured the pouches with sharpened umbrella tips. Sarin is sufficiently volatile that no special dissemination devices were necessary. Soon after the pouches were punctured, poisonous fumes filled the cars. Despite the crude technique, 12 people died and more than 5,000 were injured, many seriously enough to require hospitalization. Two subsequent attacks failed, in part because Tokyo police were on the lookout for suspicious packages. On May 5, 1995, cult members left small plastic bags, one containing sodium cyanide and the other sulfuric acid, in Tokyo's Shinjuku train station with the intention of disseminating cyanide gas. And on July 4, 1995, similar improvised chemical devices were found in restrooms in four stations.[43]

The Aum Shinrikiyo cult intended to kill many thousands of people. The 1995 poison-gas attack in the Tokyo subway was carried out in haste and did not represent the cult's full potential. The group had built up a large production capacity for chemical weapons and was working on plans for disseminating chemical and biological agents over some major Japanese city. Cult members had hidden a bottle containing an ounce of VX—reportedly enough to kill about 15,000 people—which was recovered by the Tokyo police in September 1996. They had amassed hundreds of tons of chemicals used in the production of sarin—reportedly to make enough sarin to kill millions. They had bought a Russian Mi-17 combat helicopter and two remotely piloted vehicles to disseminate the agent over populated areas. Police found a large amount of *Clostridium botulinum,* together with 160 barrels of growth media (required for growing the bacteria). The cult was reportedly cooperating with North Korea, with former Soviet Mafia groups, and indirectly with Iran, in smuggling nuclear materials and conventional munitions out of Russia through Ukraine.[44] Members reportedly visited Zaire on the pretext of providing medical assistance to victims of the Ebola virus, with the actual objective of acquiring a sample of the virus to culture as a warfare agent. The group was also actively trying to purchase Russian nuclear warheads, according to the CIA, and may have been plotting a chemical attack in the United States.[45]

In November 1995 Senator Sam Nunn held a hearing on Aum Shinrikiyo. The senator's staff testified that at the time of the Tokyo attack Aum Shinrikiyo had some 50,000 members, 30,000 of whom were Russians. It had assets worth $1.4 billion and offices in Bonn, Sri Lanka, New York, and Moscow as well as in several Japanese cities. U.S. officials admitted that, despite the alarming range of Aum Shinrikiyo's activities, it was not on the "radar screen" of the U.S. intelligence community.[46]

Other Cases

Before the Aum Shinrikiyo incidents, other terrorists had attempted or threatened to use chemical, biological, or radiological agents, but usually on a scale so small that the media hardly noticed. Most of the incidents involved threats that were never carried out, although some of the threats were very costly to the targeted companies or governments. In the rare cases when these agents have been used, they have been used more as weapons of mass impact than as weapons of mass destruction: few people have been killed.

A common use of these agents is to commit economic sabotage either against specific companies or against entire industries. For example, the Animal Liberation Front (ALF) claimed to have spiked Mars chocolate bars with rat poison to protest research on tooth decay conducted on live monkeys. No poison was found, but Mars reported losses of $4.5 million. British police later charged four members of the ALF with injecting toxic mercury into turkeys sold in supermarkets as a protest against their slaughter during the Christmas season. The same group was suspected of poisoning eggs in British supermarkets in 1989. The eggs were punctured and marked with a skull and crossbones. An attached message signed "ALF" warned that the eggs had been poisoned. While the product-tampering crimes committed in the United States in the late 1980s were not committed by terrorists, the costs imposed on industry demonstrate the potential effectiveness of economic terrorism. In 1986 alone, U.S. pharmaceutical manufacturers destroyed over $1 billion worth of pharmaceuticals because of tampering or threats of tampering and spent another $1 billion making their products more resistant to tampering.[47]

Chemical and radiological agents have also been used to meet more traditional terrorist objectives: to attack symbolic targets, to assassinate individuals, or to commit small-scale acts of random violence. Several groups have planned to attack water supplies, although no known attack has been successful. In 1972 a U.S. neo-Nazi group, the Order of the Rising Sun, was found in possession of a large quantity of typhoid bacillus, which it reportedly intended to use to poison water supplies in several midwestern cities. Another incident involved a plot by several right-wing extremist groups to overthrow the U.S. government. One of the groups, The Covenant, the Sword, and the Arm of the Lord, had stockpiled some thirty gallons of cyanide for the purpose of polluting municipal water supplies. The group was apprehended before the plot could be carried out.[48]

While the terrorists involved in these two incidents were plotting alarmingly destructive attacks, it is unlikely that either of the planned attacks would have succeeded. The chlorine in U.S. reservoirs would have killed the typhoid bacillus, and dilution would have rendered the cyanide harmless. Attacking a small, unprotected water supply could be more effective. Ramzi Youssef, the convicted mastermind of the bombing of the World Trade Center, reportedly threatened in a letter to poison water supplies in the Philippines. The letter was found on his person at the time of his arrest. In it he claimed to be able to produce chemical agents for use against "vital institutions and residential populations and the sources of drinking water."[49]

A few attacks have involved radioactive isotopes. In 1995 Shamil Basayev, the leader of the Chechen group that had earlier taken more than 1,000 hospital patients hostage, buried a packet of radioactive cesium in Izmailovski Park in Moscow to demonstrate his capabilities.[50] Izmailovski Park is a popular recreation spot for both Russians and tourists. Had Basayev actually disseminated radioactive cesium, he would have imposed heavy costs on the Russian government. In a bizarre case on Long Island in 1996, three people became convinced that county officials were covering up the crash landing of space aliens in their county. They acquired five canisters of radioactive radium, with which they planned to assassinate county officials by poisoning the officials' toothpaste, air conditioning, and automobiles. Their ultimate objective was to seize control of the county government, but members were apprehended before they were able to poison anyone, much less achieve their ambitious agenda.[51]

The single U.S. case involving the actual use of biological agents occurred in September 1984, when members of the Rajneeshee cult in Oregon poisoned salad bars with *Salmonella typhimurium.* The cult had established a commune on a large ranch, part of which was incorporated as the city of Rajneeshpuram. Local residents objected to the commune and challenged the city's charter in the courts. The cult sought to ensure the victory of its own candidate in the November 1984 elections for county commissioner. Members considered various ideas for accomplishing this goal, including vote fraud, and decided to prevent nonmembers from voting by making them ill. In September, as a trial run, members contaminated salad bars in ten local restaurants; 751 people became ill. The cult then abandoned the plot, however, in part because members judged their candidate would not prevail in the election. The Rajneeshee involvement in the outbreak of illness was not established until a year later, during investigations of cult members for other, unrelated crimes.[52]

These cases are instructive. First, no terrorist group has acquired a nuclear weapon, although Aum Shinrikiyo attempted to do so. Second, the most destructive biological attack involved a crude food poison, rather than air dissemination of a deadly agent. While such poisons are somewhat easier to procure than deadlier agents, the ability of a terrorist group to contaminate a salad bar is alarming, as is the health authorities' difficulty in determining the cause of the outbreak of illness. This case raises the question of whether other outbreaks of disease, assumed to have resulted from natural causes, may actually have been caused by deliberate sabotage or terrorism.

The most destructive incident, Aum Shinrikiyo's attack on the Tokyo subway, involved disseminating a chemical agent in an enclosed space. This method and the contamination of food are probably the easiest ways to use weapons of mass destruction, and are likely to remain the most common forms of terrorism involving these weapons.

All of Aum Shinrikiyo's large-scale attacks with biological agents are believed to have failed. Very little is known about why they were unsuccessful. There are many strains of *Clostridium botulinum* and *Bacillus anthracis.* Experts now believe the cult may have grown weak strains of the microbes, possibly, in the case of the anthrax, a relatively harmless vaccine strain. The cult also had difficulty aerosolizing the agent in respirable particle size. Sprayers used to create anthrax mists became clogged, for example, a problem that other terrorists would be likely to confront as well.[53] Were terrorists to master these technologies they would have the potential to kill not just hundreds but hundreds of thousands of people.

Who Are the Terrorists?

As the new millennium approaches, we face the very real and increasing prospect that regional aggressors, third-rate armies, terrorist groups and even religious cults will seek to wield disproportionate power by acquiring and using [weapons of mass destruction].

—Secretary of Defense William Cohen, 1997

As we have seen, technical constraints no longer prevent terrorists from using chemical, biological, and radiological agents in low-technology operations, such as poisoning foods or disseminating these agents in enclosed spaces, and constraints against higher-technology operations, including nuclear terrorism, are eroding as well. But the technical capability to employ weapons of mass destruction (WMD) does not necessarily imply that terrorists will *want* to employ them. Indeed, there are many reasons to believe they will not.

I have defined terrorism as an act or threat of violence against noncombatants, with the objective of exacting revenge, intimidating, or otherwise influencing an audience or audiences. Dread is an essential feature of terrorism: while terrorists have a variety of putative objectives, nearly all use dread as an instrument for achieving their goals.

Inspiring dread does not necessarily require killing, or threatening to kill, large numbers of people. The fear evoked by radiological, nuclear, chemical, and biological agents can be entirely independent of the number of innocent lives lost through their use. Nonetheless, many terrorists will shy away from using these weapons because they, or their constituents, consider them morally abhorrent.[1] And some other terrorists will be attracted to these weapons precisely because, when used in the right way, they have more bark than bite.

Most terrorists will probably continue to avoid WMD, for a variety of reasons. To employ WMD successfully, terrorists would have to be technically proficient, capable of overcoming moral constraints, and organized to avoid detection, and they would have to want to use the weapons despite formidable political costs. The number of groups meeting all of these requirements is likely to remain small, especially for operations in which such weapons are used to kill large numbers of people. The terrorists most likely to attempt to use WMD are groups with amorphous constituencies, including religious fanatics, groups that are seeking revenge and groups that are attracted to violence for its own sake.

Why Use WMD?

Why might terrorists, in spite of the obstacles, decide to use weapons of mass destruction? One reason might be as a way of *attracting attention.* As discussed earlier, poisons—including the radiation produced in nuclear explosions—inspire fear that is out of proportion to the actual dangers they pose. This quality may make them appealing to terrorists who feel they are not getting enough attention from a public or a government that has become jaded. Michael Baumann, the founder of a German left-wing terrorist group called the June 2 Movement, said in an interview: "During their attack on the Stockholm Embassy the RAF [Red Army Faction] people noticed that the government no longer gives in . . . Now they have something that will work for sure, and what else can that be except the ultimate thing? . . . These are intelligent people and they have vast amounts of money. They also can build a primitive nuclear bomb."[2]

Religiously motivated terrorists might decide to use WMD, particularly biological agents, in the belief that they were *emulating God.* The fifth plague with which God punishes the Pharaoh in the story of the Israelites' Exodus from Egypt is murrain, a group of cattle diseases that includes anthrax (Exodus 9: 1–6). In I Samuel 5: 9, God turns against the Philistines with a "very great destruction," killing them with a pestilence that produces "emerods in their secret parts." Medical historians believe these "emerods" to be a symptom of bubonic plague.[3] Some terrorists might feel they were following God's example by employing these agents.

The core myth for some of the white supremacist groups known as Christian Patriots is the story of Phineas, who took the law into his own hands by murdering a tribal chief and the chief's foreign-born concubine for flouting a proscription against miscegenation. Through this act of brutal murder, Phineas "purified" the community, which until then had tolerated "whoredom with the daughters of Moab," which was forbidden by God. God rewarded Phineas by averting a devastating plague: "And when Phineas, the son of Eleazar, the son of Aaron the priest, saw [the union of the Israelite and the foreign-born woman], he rose up from among the congregation, and took a javelin in his hand; And he went after the man of Israel into the tent, and thrust both of them through, the man of Israel, and the woman through her belly. So the plague was stayed from the children of Israel" (Numbers 25: 7–8). Christian Patriots use this biblical story to justify murdering mixed-race couples and nonwhite "mud people." They might eventually use it to justify punishing those "mud people" with the plagues that God averted.[4]

Another possible motivation for terrorists' use of WMD is *premillennial tension.* The millenarian idea is that the present age is corrupt and that a new age—the millennium—will dawn after a cleansing apocalypse. Only a lucky few will survive the apocalypse and experience paradise.[5] A more mundane meaning of the word *millennium* is simply a period of a thousand years, and as the year 2000 approaches and the calendar's millennium draws to an end, many millenarians expect the apocalypse and the dawning of the mystical millennium.

Judeo-Christian apocalyptic prophecies envision a series of devastating catastrophes as critical signs of the Messiah's imminent appearance. Most millenarians believe that God is responsible for determining the date of the apocalypse, which is a precondition for the appearance of the Messiah. But some believe that humans can speed the process along. Prayer, repentance, and martyrdom are common techniques—but some millenarians may add terrorism to this list.

A member of the millenarian group known as The Covenant, the Sword, and the Arm of the Lord explained in an interview why his group had planned to poison city water supplies with cyanide in the 1980s: "We thought there were signs of Armageddon, and we believed that once those signs were there it was time for us to act, to make judgments against those who were doing wrong or who refused to repent. We felt you could kill those people, that God wanted us to kill those people. The original timetable was up to God, but God could use us in creating Armageddon. That if we stepped out things might be hurried along. You get tired of waiting for what you think God is planning."[6]

Not all millenarian movements use violence. And some turn violent against themselves. The group that called itself Heaven's Gate, based in a suburb of San Diego, California, was inspired by the appearance of the Hale-Bopp comet to commit mass suicide in 1997. Thirty-nine cult members "shed their containers" in the hope that their spirits could board a spaceship they believed was hiding in the comet's tail.

A critical indicator of a group's proclivity toward violence is the nature of its core myth. Groups that model themselves on an avenging angel or a vindictive god (such as Christ with a sword, Kali, or Phineas) are more likely to lash out than those whose core myth is the suffering Messiah, though some movements switch myths under pressure (such as the pressure caused by the imminence of the millennium). Other indicators of a proclivity toward violence, according to Jean Rosenfeld, include the number of disciples granted sacred authority (a single leader is more dangerous than a group), and a failure to extend civil rights to dissidents during conflicts.[7] Millenarian organizations operating during periods when apocalyptic fears are common are more likely to become violent.

The Christian Research Institute, which tracks what it calls "millennial madness cults," reports that thousands of these groups are now in existence. As we approach the end of the millennium, the director of the Institute says, millenarian and Christian groups that believe in UFOs are increasing in number, and many of them are using the Internet as a form of "techno-shamanism" to create "mystical connections" in cyberspace.[8]

Many seers have predicted that the world will end around the year 2000, and some 350 organizations subscribe to this view, according to a recent study. Prophets who have foreseen the world's demise include Nostradamus, Edgar Cayce (a famous American clairvoyant), Sun Bear (a Chippewa medicine man), Alice Bailey (an English mystic), Our Lady of Fatima, the twelfth-century Irish seer Saint Malachy, and the Seventh-day Adventists. But the seers, Walter Laquer notes, do not agree on the exact day the world will end: "Some have opted for the 1990s, others for 1999, yet others for 2000 or 2001; to be on the safe side, some give a vaguer, later date." Nor is there consensus on the exact circumstances or location of the apocalypse. Some foresee floods or conflagration, others epidemics such as AIDS and starvation, while others predict that wars or nuclear terrorism will bring about the end of the world. What is clear is that as the chronological millennium approaches millenarians are likely to become more dangerous—both to others and to themselves.[9]

Terrorists might also seek out WMD to commit *economic terrorism*. Unlike conventional weapons, radiological, chemical, and biological agents could be used to destroy crops, poison foods, or contaminate pharmaceuticals. They could also be used to kill livestock. Such acts could inflict enormous economic and political costs on the target country, industry, or company. Terrorists might use these agents to attack corporations perceived to be icons of the target country—for example by contaminating batches of Coca-Cola, Stolichnaya vodka, or Guinness stout. Terrorists could also employ threats (as distinct from actual use) of these agents to force the government to recall foods or pharmaceuticals.

Some terrorists might choose WMD to impress their target audiences with *high technology*. Terrorists find technology appealing for a variety of reasons. Shoko Asahara, the leader of the Aum Shinrikiyo cult, believed technology would give him an edge over aliens, to whom he attributed "levels of consciousness" higher than that of human beings. "They may be from the Heaven of Degenerated Consciousness or the Heaven of Playful Degeneration. Anyhow, they are superior to human beings." But aliens are not to be trusted: "I hear these aliens eat human soup . . . just like humans eating beef." The only hope for humans, Asahara warned, was to surpass the aliens in technological prowess.[10]

The leader of the neo-Nazi group National Alliance, William Pierce, who studied physics at the California Institute of Technology, is also interested in high-technology weapons. In his novel *The Turner Diaries,* right-wing extremists use nuclear, chemical, biological, and radiological weapons to take over the world. Pierce believes he can attract more intelligent members to the Alliance over the Internet than over the radio or through leaflets. He is explicitly attempting to recruit military personnel, presumably because they have access to higher-technology weapons than those available on the street.[11]

Neo-Nazi groups and other admirers of Hitler may turn to nerve agents because of their use by the original Nazis, as a way to *emulate Hitler.* Aum Shinrikiyo cult members found the Nazi origins of nerve gas appealing. Japanese authorities found a manual at an Aum facility that contained, in addition to the chemical formula for sarin, a poem called "Song of Sarin":

> It came from Nazi Germany, a little dangerous chemical weapon,
> Sarin, Sarin!
> If you inhale the mysterious vapor, you will fall with bloody vomit
> from your mouth,
> Sarin, sarin, sarin, the chemical weapon!
> Song of Sarin the Brave . . .[12]

When it comes to methods, terrorists tend to *copy others,* so the use of WMD by one terrorist group or by a nation may spark similar incidents. As a rule, terrorists stick to proven tactics. Occasionally they develop new techniques or technologies, sometimes creating new norms.[13] Until the 1960s assassination predominated as a terrorist tactic. In the late 1960s terrorists crossed a moral threshold: they began to launch random attacks, killing innocent people who just happened to be in the wrong place at the wrong time. Hostage taking became popular by the early 1970s. Hijacking planes and attacking embassies emerged as tactics in the late 1970s. A norm of sparing women and children began to erode in the mid-1980s, when several groups began deliberately to target women and children. For example, in 1989 the

Provisional Irish Republican Army (PIRA) assassinated the German wife of a British soldier and murdered another British soldier and his child. A rash of car bombings followed the first few that occurred in Lebanon in the early 1980s. In 1985 Sikh terrorists initiated the practice of blowing up airplanes, killing 363 people on an Air India Jet.[14]

The CIA, in its 1984 "Special National Intelligence Estimate," reportedly claimed that chemical and biological agents were not yet popular among terrorists, probably because the terrorists themselves were terrified of the weapons, but that one successful incident involving these weapons "would significantly lower the threshold of restraint on their application by other terrorists." Aum Shinrikiyo's gas attack in the Tokyo subway was inspired by Iraq's use of chemical weapons during its war with Iran, according to Brian Jenkins. And many analysts fear that Aum Shinrikiyo's attack will inspire additional, similar use of chemical agents.[15]

Terrorists could force the government to *evacuate a city,* and to engage in a costly clean-up operation, by using radiological agents or anthrax spores. Anthrax would kill the city's residents, but the terrorists could warn authorities in advance if evacuation, not killing, was their goal. Unlike conventional weapons, anthrax and radiological agents could persist in soil or structures for many years.

As governments implement more sophisticated security measures against terrorist attacks, terrorists might find WMD appealing as a way to *overcome such countermeasures.* When governments improved their methods of protecting embassies and preventing hijacking, terrorists were forced to contrive new techniques. Metal detectors made it harder to carry guns onto planes, so rather than hijacking planes, terrorists began blowing them up with plastic explosives. Concrete barriers at U.S. embassies and government buildings made it more difficult to drive cars or trucks full of explosives onto the sites, so terrorists started using more powerful explosives. If airports routinely deploy plastic-explosive detectors, terrorists may, for example, attempt to disseminate chemical or biological agents on planes.

Terrorists whose constituents have been targeted with WMD might acquire them to *retaliate in kind.* After Iraq's brutal use of chemical weapons against the Iranian-held town of Halabja in 1988, a group calling itself the United Organization of the Halabja Martyrs claimed credit for an attack at the British Club in Iraq that injured twenty people in 1989.[16] While this group did not employ chemical agents, it is not hard to imagine terrorists turning to these weapons in, say, the former Yugoslavia, where various factions have accused one another of poison-gas attacks.

Finally, some terrorists may be drawn to WMD because their goal is to kill many people—to commit *macroterrorism.* In the past, in Brian Jenkins's famous words, terrorists seemed to want "a lot of people watching, not a lot of people dead." Terrorists aim to harass, not to kill in large numbers, Kenneth Waltz argues.[17] But it is increasingly clear that not all terrorists feel that way.

The ad hoc group of radical Islamic fundamentalists responsible for the World Trade Center bombing in 1993 intended to bring the World Trade Center buildings down. Had they not made a minor error in the placement of the bomb, the FBI estimates, some 50,000 people would have died. Investigators found a bottle of sodium cyanide in the bombers' warehouse. Some experts are convinced that cyanide was used in the bomb, but that it burned instead of vaporizing. It is unclear, however, whether the bomb actually contained cyanide or whether the terrorists had acquired the poison to use in other possible attacks.[18]

Also in 1993, the FBI thwarted a "summer of mayhem" planned by another multinational group of radical Islamic fundamentalists. The plotters were in the act of mixing explosives when the FBI, which had an informant within the group, moved in and arrested eight suspects. The plan was to bomb the United Nations building, the Lincoln and Holland tunnels, and the Federal Plaza; and also to assassinate the President of Egypt, the Secretary General of the United Nations, and two members of the U.S. Congress.[19]

These groups clearly intended to kill large numbers of people. They failed in the first case because of a minor technical error and in the second because the FBI had been informed of the plot. Terrorists whose goal is to kill many thousands might try to disseminate biological agents over a city, or (less likely) to acquire a nuclear weapon. Ad hoc groups seeking revenge are possible candidates for this type of terrorism.

Technical Obstacles

Some kinds of terrorist groups are more likely than others to be able to surmount the technical constraints on the use of WMD (described in Chapter 4). Fortunately, there is likely to be a negative correlation between psychological motivation to commit extremely violent acts and the actual ability to do so. Schizophrenics and sociopaths, for example, may *want* to commit acts of mass destruction, but they are less likely than others to succeed.[20] Schizophrenics, in particular, often have difficulty functioning in groups, and group effort would be necessary for large-scale dissemination of chemical, biological, or radiological agents, or for producing a nuclear device. But the Internet may make it easier for individuals who dislike groups to communicate with others and to collect weapons-related information with minimal face-to-face contact.

State-sponsored terrorists are the most likely to be able to overcome technical obstacles to the use of WMD, but the threat of retaliation against the sponsoring states may deter the states from supplying the weapons. Nonetheless, a number of violent incidents in recent years have made clear that it would be imprudent to rely exclusively on traditional deterrence to prevent state-sponsored acts of WMD terrorism: some delivery systems have no return address. One particularly worrying state sponsor is Iraq (see Chapter 7).

Failure is more likely if the weapon contemplated is highly sophisticated, such as a nuclear device or a "superplague" capable of killing hundreds of thousands. As noted earlier, however, low-technology missions, such as dissemination of chemical, biological, or radiological agents within enclosed spaces, would be relatively easy to carry out even without state sponsors. Probable indicators of a group's ability to overcome technical hurdles include previous use of high-technology weapons; state sponsorship; access to financial resources; a relatively large, well-educated membership; and links with corrupt government officials, weapons scientists, or organized crime.

Attempts to acquire WMD could make a group more vulnerable to detection by authorities. Groups might try to evade detection by choosing low-technology, small-scale operations; producing as many components as possible themselves; acquiring ingredients of the weapons gradually, for example by stealing small quantities of nuclear materials over a long period; using old-fashioned methods of manufacture long ago rejected as inefficient by military establishments; and assembling the weapons immediately before they are to be used.

Political Obstacles

Terrorists hoping to influence policy will presumably avoid WMD because their use would be counterproductive. These groups will not wish to alienate their constituents. For example, although some segments of the public were sympathetic to the objectives of left-wing militants in the United States in the 1960s, support began to erode as their actions became more violent. The broader radical political movement was in decline, and the militants misjudged their audience.[21]

Groups with amorphous constituencies are less likely to worry about the moral approbation of their public. Thus ad hoc groups bent on revenge, such as those affiliated with Osama bin Laden, are probably more likely to use WMD terrorism than the IRA, which is dependent on its constituency for financial support. But the question remains whether groups with clearly defined constituencies will engage in low-level violence employing WMD.

This reasoning assumes that the terrorists are actually trying to achieve their purported political goals, and that they are capable of rational analysis. But if terrorists are calculating costs and benefits, it is hard to understand why politically motivated terrorism persists. As Thomas Schelling has wryly observed:

> Acts of terrorism almost never appear to accomplish anything politically significant. True, an intermediate means toward political objectives could be attracting attention and publicizing a grievance, and terrorism surely attracts attention and publicizes grievances. But with a few exceptions it is hard to see that the attention and the publicity have been of much value except as ends in themselves . . . Despite the high ratio of damage and grief to the resources required for a terrorist act, terrorism has proved to be a remarkably ineffectual means to accomplishing anything.[22]

Some terrorists become more interested in maintaining the integrity of their group than in achieving their purported goals, so a strategy that looks irrational may in fact be rational if we understand their true objectives. And terrorists pursuing revenge, without any particular political objective, are unlikely to be concerned about political costs. For them it may be rational to seek out the most horrifying, and possibly the most destructive, weapon available.

Also unlikely to be deterred by fear of public disapproval are groups who seek chaos as an end in itself. William Pierce of the National Alliance believes that chaos may ultimately be good for American society: "This society is in a process of self-destruction. I see no signs, cannot see any conditions that will halt this process. I think I understand why this society is coming apart. It is not possible that the government or any institution will reverse this trend. Society will descend into chaos or civil war. And speeding up that process is in the interest of the country."[23]

If one political obstacle to the use of WMD is the danger of losing the support of their constituencies, another is the possibility that the government will respond with drastic measures. But some terrorists may deliberately provoke a strong government reaction in the hope of turning their constituents against the government. William Pierce believes that when the timing is right Nazi groups will be able to manipulate the U.S. government in this way, and that doing so will win them additional support. In Pierce's novel, *The Turner Diaries,* a terrorist group goads the government into taking harsh measures. The terrorists believe that by revoking civil liberties the government inadvertently reveals its own vulnerability and, by contrast, the terrorists' strength. The group adopts a strategy of continuously raising its level of violence as a way to win more sympathy and to attract new recruits.[24]

Terrorists who engage in especially provocative missions (including WMD terrorism) might elect not to claim responsibility for those acts. Anonymity would serve three purposes: it would protect the group from retaliation or arrest, it would prevent public backlash against the group, and it would increase social chaos.

Those least likely to be deterred by the risk of a government crackdown include ad hoc groups, religious zealots, and right-wing extremists seeking to overthrow the government. Because membership in an ad hoc group is fluid and temporary, the group (if not individual members) is likely to be beyond the reach of the government. And worldly consequences are not a central concern for religious terrorists, since they believe their actions are dictated by a divine authority.

Moral Obstacles

For a terrorist group to be able to use weapons of mass destruction, its members must be morally disengaged from the consequences of their actions. Albert Bandura has described four techniques of moral disengagement that are employed not only by terrorists and perpetrators of genocide but also, on a smaller scale, by decent people seeking to justify activities that further their own interests at the expense of those of others.[25]

The first technique is *moral justification.* Terrorists may imagine themselves as the saviors of a constituency threatened by a great evil, as, for example, Islamic extremists do when they label the United States the Great Satan. Similarly, The Order (also known as the Bruder Schweigen and the White American Bastion), an extremely violent right-wing group modeled on a fictional group in *The Turner Diaries,* describes its members as "soldiers" who are fighting a "Just War" against the "parasites," and claims to follow the rules of the Geneva Convention. "We now close this Declaration with an open letter to Congress and our signatures confirming our intent to do battle. Let friend and foe alike be made aware. This is war!"[26]

Donatella della Porta, who interviewed members of left-wing militant groups in Italy and Germany, observed that the militants "began to perceive themselves as members of a heroic community of generous people fighting a war against 'evil.' " This sense of heroism and righteousness sometimes persists long after the militants have achieved their ostensible goals.[27]

A second technique is *displacement of responsibility* onto the leader or other members of the group. The terrorists reconstrue themselves as functionaries or bit players who merely follow the leader's orders—a technique widely practiced in Nazi Germany. Or the terrorists may blame other members of the group. Group decisionmaking is a common bureaucratic practice that lends itself to harsh action: when everyone is responsible, no single person has to take the blame. "Social organizations," Bandura observes, "go to great lengths to devise sophisticated mechanisms for obscuring responsibility for decisions that will adversely affect others." Groups that are split into cells and columns may be more capable of carrying out ruthless operations because of the potential for displacement of responsibility. Della Porta's interviews with left-wing militants suggest that the more compartmentalized a group is, the more it begins to lose touch with reality, including the actual impact of its own actions.[28]

Some groups blame the government for acts they themselves have committed or intend to commit in the future. Aum Shinrikiyo accused the CIA of actions that the cult itself was planning, such as using chemical agents against the Japanese people. The Christian Identity news media began warning of a "biological genocide" plotted by the U.S. government at about the same time that several right-wing extremists were found in possession of biological agents (in 1995).[29]

A third technique is to minimize or *ignore the actual suffering of the victims.* The use of time bombs, biological weapons (because of the incubation period), or long-range weapons facilitates such a strategy. In a study of the psychology of killing, Dave Grossman found that the trauma of killing is highly correlated with the killer's proximity to the victim.[30]

The fourth technique is to dehumanize victims as "mud people," "the infidel," or subhumans. Neo-Nazi hate groups employ this technique in their speeches and writings. Nonwhites are called the "children of darkness" and Jews the "destroying virus," while Aryans are described as "pure," "the Chosen," and the "children of light."[31]

Most people are reluctant to kill, even during war. During World War II, Grossman reports, 75 to 80 percent of riflemen refused to fire at an exposed enemy—even to save their lives or the lives of their compatriots. But people can be trained to ignore the suffering of their victims. In Vietnam, after extensive "desensitivity training," only 5 percent refused to fire. Terrorists too can be trained to kill. As terrorists' assignments gradually become more violent, their capacity for moral revulsion is worn down.[32]

The Thugs, a violent group that operated in India for centuries (see Chapter 2), were trained as killers beginning in childhood. In response to a British interrogator's question, one Thug said he felt no more regret about having murdered people than his interrogator felt about hunting game. Stalking men "is a higher form of sport," he explained, "for you *sahib* have but the instincts of wild beasts to overcome, whereas the Thug has to subdue the suspicions and fear of intelligent men."[33]

Italian and German militants justified violence by depersonalizing their victims as "tools of the system," "pigs," or "watch dogs." One German militant said in an interview: "What I liked in that activity was the fact that it was possible, in some moments, to overcome the inhibitory restraints . . . Even today, I do not feel any general scruple concerning a murder, because I cannot see some creatures . . . as human beings."[34]

Certain types of terrorist groups are more likely to be capable of disengaging morally. Closed cells may be less susceptible to outside pressures, and members may gradually see themselves as separate from society and not subject to its norms. Groups separated into cells and columns or phantom cells may blame outside authorities for directing or inspiring acts of cruelty. Racial supremacists and religious extremists often see members of other races or religious groups as subhuman. And followers of charismatic leaders are likely to feel that the leaders are responsible for group actions and they themselves are just following orders.

Organizational Factors

When individuals come together in groups of any kind—executive task forces, religious groups, university departments—they become susceptible to group dynamics, including a tendency to engage in what has been called "groupthink," which is characterized in part by an inability to tolerate differences of opinion among members. Psychologists have also found that groups may develop unrealistic beliefs in their own morality and invulnerability, leading them under some circumstances to take more risks than individuals.[35]

Group dynamics exerts a particularly powerful influence when groups operate outside the law and are forced to go underground, since members are forced to rely exclusively on one another. In such circumstances the survival of the group—rather than any particular political objective—becomes the primary goal.

In interviews with Argentine guerrillas, Maria Moyano found that although the guerrillas initially pursued political objectives (fighting military rule), they continued to operate even after the country's return to constitutional rule. Over time they became militarized and increasingly violent, apparently pursuing violence for its own sake. Della Porta's interviews with left-wing European militants show that as these organizations cut off ties with outsiders, commitment to the cause becomes self-generating, and risk-taking increases.[36]

Thus terrorists might employ WMD, not to pursue political objectives, but to maintain the integrity of the group or to meet their own psychological needs. They might be especially prone to acts of extreme violence if they fear their group is in jeopardy. This hypothesis is consistent with prospect theory (see Chapter 3), which suggests that people are more likely to take extreme risks when facing grave losses. Terrorists might resort to more lethal tactics to avenge the deaths of members of their group or to retaliate against perceived slights.[37]

A group's chances of maintaining its own cohesion while planning to use weapons of mass destruction will depend in part on the nature of its leader. The more charismatic and powerful the leader, the greater the likelihood of success.

Leaders also need to prevent law enforcement authorities from penetrating the group. A group's location (terrorists based in remote areas, like the Unabomber in Montana, are less likely to attract attention), and its ability to recruit law-enforcement officials to the cause (as Aum Shinrikiyo and the Branch Davidians did) are important factors in this regard.

Structure is also a factor. Ad hoc groups that come together to carry out a single operation, such as the group responsible for bombing the World Trade Center, are difficult for authorities to detect and infiltrate, since they exist for a limited period of time.

The phantom cell structure advocated by right-wing extremists in the United States is explicitly designed to minimize the risk of FBI penetration. Louis Beam, who formerly served as the Grand Dragon of the Knights of the Ku Klux Klan, is described in Aryan Nations literature as their ambassador-at-large, staff propagandist, and "Computer Terrorist to the Chosen." Beam openly urges white supremacists to form phantom cells of between one and twelve members in a strategy of "leaderless resistance." These cells are encouraged to act alone—without communicating directly with the leadership of the movement:

> In the pyramid form of organization, an infiltrator can destroy anything which is beneath his level of infiltration, and often those above him as well. If the traitor has infiltrated at the top, then the entire organization from the top down is compromised and may be traduced at will . . . Utilizing the Leaderless Resistance concept, all individuals and groups operate independently of each other, and never report to a central headquarters or single leader for direction or instruction, as would those who belong to a typical pyramid organization.[38]

Timothy McVeigh acted in the way prescribed by Louis Beam as he planned and carried out the Oklahoma City bombing.

A terrorist group will find it easier to carry out some actions, such as the dissemination of chemical agents, if its members are willing to lose their own lives in the process. Suicidal missions have become common in the Middle East. Islam, like Christianity and Judaism, forbids suicide. To be killed in holy war (Jihad) guarantees paradise after death, but to kill oneself is strictly forbidden. The accepted theological view is that a person who commits suicide is destined for hell, where he will be doomed forever to repeat the act of killing himself.[39]

But Hezbollah's religious leaders sought to justify the series of suicide bombings of the mid-1980s, for which Islamic Jihad claimed credit. Ayatollah Sayyid Muhammad Fadlallah (often identified as the spiritual leader of Hezbollah although he denies any formal connection) justifies suicide according to a convoluted logic, explained by Martin Kramer. First, when there is a severe imbalance in forces, the victim of oppression discovers "new weapons and new strength every day." The oppressed must fight with special means of their own, including unconventional or seemingly primitive means of warfare. Although

Fadlallah denies he has told anyone to "blow yourself up," he pronounces that "Muslims believe that you struggle by transforming yourself into a living bomb like you struggle with a gun in your hand. There is no difference between dying with a gun in your hand or exploding yourself." There is no moral difference, Fadlallah argues, "between setting out for battle knowing you will die *after* killing ten, and setting out to the field to kill ten and knowing you will die *while* killing them."[40]

But how do terrorist groups get their members to die for the cause? Ariel Merari describes three ways. One technique, apparently practiced in the Middle East, is to force or trick people to conduct suicidal missions against their will. Merari describes several cases, but there may be more: it is impossible to question an apparent suicide about his true motivations unless the operation fails and he survives. In one case the Israeli police captured a sixteen-year-old who was about to carry out a suicide bombing. The boy claimed that he had no desire to kill himself—for Allah or anyone else—but that the Shi'ite militia had compelled him to carry out their orders by threatening his family. A second technique is to seek out mentally disturbed persons. After the series of suicide attacks in Lebanon in 1983, there were rumors among Shi'ites that the leadership of Islamic Jihad had found mentally unbalanced youths to carry out the missions, taking advantage of their psychological vulnerability. A third technique is to indoctrinate or brainwash recruits to believe that dying in the service of their leader is an honor, as kamikaze pilots seem to have believed.[41]

Terrorists have long been capable of more lethal acts of violence than they have actually committed, suggesting that they have not wanted to kill large numbers of people. Politically motivated terrorist groups are likely to avoid large-scale use of weapons of mass destruction, for fear of alienating their constituents or evoking harsh reactions from authorities. But these constraints will not apply if a group is pursuing chaos, if it has an amorphous constituency, or if it is confident of its ability to remain anonymous or evade law enforcement.

While state-sponsored groups are the most technically proficient, fear of retribution is likely to play a role in their decisions about whether to use weapons of mass destruction. Two kinds of groups are of particular concern: millenarian groups, including Christian Patriots, and radical Islamic fundamentalists organized as ad hoc groups. Christian Patriots are especially troubling because they are often organized to evade detection by authorities, they are interested in WMD, and they have demonstrated an ability to acquire chemical and biological agents. Some Islamic extremist groups, such as those affiliated with bin Laden, also appear to be capable of overcoming all four obstacles—technical, political, moral, and organizational—to the use of WMD, and to be strongly motivated to use them.

The Threat of Loose Nukes

The question is not whether large quantities of highly enriched nuclear material will be stolen, but when.

—Russian Ministry of Atomic Energy official, 1994

In May 1993 Joseph Rimkevicius, chief of the organized-crime control section of the Lithuanian police, received an anonymous tip that someone was storing valuable—but unspecified—smuggled goods in the Joint-Stock Innovation Bank in Vilnius. Ten men had already died in Klaipeda, a port city in western Lithuania, in a fight between rival organized-crime groups over the mysterious cache. Several days after receiving the tip the Lithuanian police staged a bomb threat as a pretext for inspecting the Innovation Bank. The detectives found twenty-seven crates holding thousands of metallic cylinders. The cylinders turned out to be composed principally of beryllium, a very light metal. The total amount discovered was astonishingly large: 4,000 kilograms of beryllium, which sells on the world market for around $600 a kilogram, stored in two Lithuanian banks. Interpol later learned that a buyer in Switzerland had offered $24 million for the entire cache, about ten times the market rate.[1]

Beryllium is a very low-density metal—lighter than aluminum, but stronger than steel. It is used in inertial guidance systems for missiles, in high-performance military aircraft, and in optical equipment. And beryllium is a neutron reflector. It is this property that makes beryllium useful in nuclear power plants as well as in the production of nuclear weapons. If a shell of beryllium is placed around a sphere of fissile material, neutrons created during the fission reaction are reflected back into the sphere, enhancing the chain reaction. This allows the bomb designer to use a smaller quantity of fissile material for a given yield, thereby reducing the weight of the bomb.

A lengthy investigation revealed that the source of the beryllium, which was contaminated with small amounts of HEU, was the Institute of Physics and Power Engineering in the Russian city of Obninsk, sixty miles from Moscow.[2] Authorities discovered that a company in Ekaterinburg with links to organized crime had purchased the beryllium from Obninsk, and that senior Russian government officials had signed off on the illegal transaction. Like Chechnya, Ekaterinburg is a leading center of organized crime in Russia. The Baltic states have particularly poorly guarded borders, and organized crime is rampant there as well. Thus it is not surprising that the exporters chose to smuggle the nuclear-related material through Lithuania, as its borders were almost entirely unguarded.[3]

The most profound contribution to the increased danger that terrorists will acquire and use weapons of mass destruction (WMD) is the deeply chaotic condition of the former Soviet Union and its lack of control of its nuclear and chemical arsenals. The most obvious symptoms of chaos are corrupt generals selling weapons abroad, inadequately policed borders, and poorly secured facilities housing weapons and their components. The most significant threats to U.S. national security now arise not from Russia's military might but from its weakness.

To date most of the nuclear-smuggling incidents that have come to authorities' attention have been perpetrated by amateurs at former Soviet nuclear sites. Typically the material is stolen by workers or other insiders who have heard that nuclear materials fetch high prices on the black market, but who have no experience in smuggling illicit materials and no idea how to locate potential buyers. In some cases the thieves have hidden the stolen material for months while searching for buyers. Selling stolen nuclear material has something in common with selling stolen art: the risks of being caught impede buyers and sellers from finding each other. Some thieves have been confused about the difference between medical isotopes and material that can be used to make nuclear weapons, and some have caused injury by exposing themselves or others to radiation. A 1996 task force composed principally of U.S. officials concluded that organized crime has been involved in the trafficking of radioactive materials, but that "the weapons-usable material trade appears to be low-level and localized."[4] Although the beryllium case involved very little material usable for nuclear weapons, it was unusual—and unusually troubling—because of the confirmed involvement of high-level government officials and organized crime.

Seizures of stolen nuclear materials by the authorities appear to have peaked in 1994, but it is unlikely that smuggling has actually decreased. Rather, as suggested by Vladimir Orlov, a Russian expert on the security of nuclear materials, it may be that the amateur thieves have become more proficient, that professionals have entered the market, or that the Russian government has become more secretive.[5]

The real threat of nuclear smuggling is not what we have seen so far. Corrupt government officials and intelligence agents are in a much better position than amateur thieves to know how to export illicit materials, especially if they combine forces with organized crime. Organized-crime groups are now serving, according to a senior U.S. intelligence official, as "intermediaries between senior government officials": "They have enough money to carry off diversion of weapons of mass destruction. The threat is no longer hypothetical." Several criminal organizations have connections at the highest levels of governments throughout the former Soviet Union and elsewhere. They include among their members former intelligence operatives. And they have demonstrated an ability to move large quantities of metals and other goods across borders illegally; this same expertise might be used to export large quantities of fissile material or even bombs. For example, Nordex, a large metals-trading firm based in Austria, was allegedly founded with KGB funds, and some experts suspect that Nordex was involved in the beryllium case.[6]

Military Meltdown

The Russian military is in crisis. Living conditions are abysmal, food and housing are in short supply, and wages are often delayed.[7] More than 500 officers committed suicide in 1996. The chief military procurator claims that embezzlement by officers is common. There is so much theft of military equipment that, one official complains, "the country is awash with arms, nobody knows how many." Deborah Ball, who surveyed Russian officers in 1995, believes that the officer corps can no longer be relied on to obey lawful orders during political disputes. Even the elite Strategic Rocket Forces are barely eking out a living, according to General Lebed, who once headed President Yeltsin's Security Council.[8]

Many Russian and U.S. officials are more sanguine about physical security for warheads than for nuclear materials. There is a basic "guards, gates, and guns," approach to warhead security, a U.S. official told me. "You can't put [warheads] under your overcoat. It's clearly much harder to steal a warhead than to steal the materials to make one." Senior U.S. officials continue to claim that "despite the general breakdown in the Russian military caused by severe economic hardship, the approximately 6,000 strategic and 22,000 tactical nuclear arms at about 60 sites remain safe and under tight government control."[9] But it is clear that Russia's transition from an authoritarian state to a struggling, chaotic democracy is subjecting the nuclear security system to stresses that were not envisioned when the system was developed.[10]

General Lebed told a visiting U.S. congressional delegation in May 1997 that of 132 "suitcase bombs" (small nuclear weapons called atomic demolition munitions) in the former Soviet arsenal, he had been able to locate only 48, leaving 84 unaccounted for. Later Lebed retracted his claim, but later still repeated it, on the BBC in November 1997, this time providing the name of the devices (RA-115), their weight (30 kilograms) and their yield (2 kilotons).[11]

Other senior Russian officials have also told congressional delegations that tactical nuclear weapons are missing. Because tactical weapons are small and hard to track, it is difficult to refute—or confirm—such claims, which may be inspired by Russian domestic political considerations.[12] In July 1997 two Lithuanian citizens were arrested for conspiring to "transfer and disperse without lawful authority nuclear materials," including tactical nuclear weapons, to U.S. federal agents posing as drug smugglers. Although the federal agents never saw the weapons, the accused men supplied them with "enough information to convince them that they could make good on their word." The Lithuanian government disputed the U.S. agents' claims, calling the two immigrants petty swindlers incapable of acquiring nuclear weapons.[13]

The system of command and control for nuclear weapons is also degrading. In January 1995 Russian radar detected an incoming missile, which the command system interpreted as a possible attack by the West. For the first time in its history, the command system started a countdown to launch nuclear weapons. President Yeltsin's nuclear "football" was activated; an emergency teleconference was triggered.[14] After eight minutes—only a couple of minutes before the procedural deadline for a decision to launch—authorities realized it was a false alarm. The missile was in fact a U.S. scientific rocket, which had been launched from Norway to study the northern lights. Bruce Blair warns that the "progressive deterioration of Russian command-control and early warning networks represents the most serious current threat to Western security . . . It is not unreasonable to anticipate a catastrophic failure of Russian nuclear control if current trends persist."[15]

The CIA, in a leaked September 1996 report, said that "control over Russia's vast nuclear arsenal is growing weaker, and a political crisis could lead to an unauthorized strategic missile attack by renegade military officers." Not surprisingly, the Russian government denied the report. The CIA cited Russian officials' concerns about nuclear units in the Far East, "where troop living conditions are particularly deplorable and where nuclear weapons might fall into the wrong hands." All blocking devices—on both strategic and tactical weapons—can be defeated given enough time. And "some submarine crews probably have an autonomous launch capability for tactical nuclear weapons and might have the ability to employ SLBMs [submarine-launched ballistic missiles] as well." While the CIA concluded that under "normal circumstances," unauthorized launch remained unlikely, it noted that a severe political crisis would make nuclear blackmail or unauthorized launch more likely.[16] Experts outside the government have expressed similar views.[17]

With rare exceptions, in public statements Russian officials have denied that nuclear warheads are vulnerable to terrorist attack or to theft. But the government has nonetheless taken steps to remove nuclear weapons from the volatile Caucasus region and to consolidate nuclear warheads in storage—from more than 600 sites in 1989, to 200 sites in 1991, to fewer than 100 in 1995. And some officials have admitted concerns about the security of the warheads in transit. "What is theoretically possible and what we must always be prepared [for] is train robbery, attempts to seize nuclear weapons in transit," a senior Russian military official has admitted. "We ran some modeling exercises at our facilities [to test our warhead security system] . . . And I must tell you frankly that as a result of those exercises, I became greatly concerned about a question that we had never even thought of before: What if such acts were to be undertaken by people who have worked with nuclear weapons in the past? For example, by people dismissed from our structures, social malcontents, embittered individuals?"[18]

In an interview in November 1995, the official assured me that an inventory is taken of all Russian warheads twice a year, in which the seals are removed. But a Ministry of Atomic Energy official filled in some disturbing details. Seals are removed to assess the electronic equipment inside the warhead, not to verify the presence of nuclear material. One could easily replace a warhead with an "imitator." The substitution would not be noticed for many months because the seals are of poor quality and can be "falsified." The Ministry of Defense does not understand what it means to conduct an inventory, the official told me. It uses a paper-based system vulnerable to human error, in which warheads are counted *na pal'tsyax* (with the fingers). The system was not designed, he said, to deter threats from insiders.[19]

The Nuclear Economy under Threat

It is only now becoming clear to what extent military production dominated the Soviet Union's economy. It is very difficult to calculate the percentage of the Soviet GNP accounted for by defense-related production—even the GNP itself is nearly impossible to assess—but estimates now range between 12 and 25 percent, an extraordinarily high level of defense expenditures. The fraction of the U.S. GNP devoted to defense, in contrast, hovered around 6.5 percent during the 1980s.[20]

A significant part of the Soviet defense-industrial base included facilities to produce or process nuclear materials—both for warheads and for reactor fuel. There are around forty such facilities in Russia today, and a dozen or so in the other former Soviet states. Ten of these facilities are located in "closed cities" exclusively devoted to secret military nuclear research and production. In the past the 725,000 citizens living in closed cities had little contact with the outside world. Now Russia's problems, including unemployment and organized crime, have entered the closed cities' gates.[21]

Spending on weapons research has declined by as much as 70 percent. Once treated as the elite, many scientists in the nuclear weapons industry are now poverty-stricken. Workers and scientists often go unpaid for months at a stretch, a situation that leads to social unrest and increases the risk of accidents. Real wages are below subsistence level for employees at Arzamas-16, the laboratory where Russia's nuclear weapons were designed, employees who are "literally holding a nuclear weapon in their hands."[22]

Former weapons scientists are searching for alternative ways to feed their families. Some are now working at menial jobs, barely making a living. And some are reportedly supplementing their incomes by providing consulting services to Iran and Pakistan on weapons-related research. American officials suspect that the Russian government is aware that Russian scientists have been supporting Iran's efforts to develop long-range missiles, despite the Russian government's protestations to the contrary.[23] Russian experts fear that weapons scientists may become so desperate they will sell nuclear secrets or nuclear materials abroad. In late 1996 the director of Chelyabinsk-70, one of Russia's most elite nuclear weapons laboratories, killed himself, claiming he could no longer bear his inability to pay his workers.[24]

The Ministry of Atomic Energy has informed the managers of nuclear facilities that they will have to meet their own operating costs by marketing goods and services. In June 1997 Viktor Mikhailov, who was then the Minister of Atomic Energy, announced a plan to double exports of nuclear materials and technology by the year 2000. The Ministry's exports of nuclear technology for 1996 were worth more than $2 billion, a 20 percent increase over 1995. Since the Ministry's facilities are allowed to keep an undisclosed percentage of its exports, and since many of these facilities are experiencing budget shortfalls, nuclear facilities face strong incentives to maximize exports, even if some of those exports might fall into dangerous hands. The trend is not good: the Ministry called 1997 its "worst ever" year in terms of financing.[25]

The Ministry of Atomic Energy is concluding deals to build a nuclear power station in India and to complete construction of the Juragua plant in Cuba, and is going forward with construction of a 1,000-megawatt reactor in Bushehr, Iran, despite U.S. opposition. A secret protocol with Iran—which included provisions to supply research reactors, to train Iranian physicists, and, most troubling, to provide technology for uranium enrichment (necessary to create HEU)—was reportedly canceled in 1995. The Ministry appears to have negotiated the protocol without consulting the presidential staff or other Russian agencies: the Ministry of Foreign Affairs seemed to be as surprised by the secret protocol as was the U.S. government.[26]

There is growing sentiment in Moscow that the Ministry of Atomic Energy is functioning as a "state within a state," and should be reined in. For example, in February 1996 Minister Mikhailov announced that Russia would preemptively attack any tactical nuclear weapons deployed in former Warsaw Pact countries that hope to join NATO.[27] This statement was later retracted by other officials. The root of the problem, said a Russian ambassador who asked me not to use his name, is that Stalin was in such a rush to develop the bomb that he gave the Ministry of Atomic Energy extraordinary powers, including its own cities where workers received special privileges in exchange for working under extreme secrecy. The Ministry had to function as the state in those cities, and it grew accustomed to power and secrecy. The Ministry also employs more than a million workers, who represent a sizable voting block.

Inadequate Security

The Soviet system for protecting nuclear materials was not designed for a democratic state, an official from Russia's Ministry of Atomic Energy told me. It was designed with two objectives: to prevent attacks by outsiders, such as terrorists; and to keep American spies from acquiring nuclear secrets. No one considered the possibility that workers themselves might steal nuclear materials. The system was based on "regulations and ordinances which either no longer are in place or not effective, and upon military discipline and sense of responsibility which no longer exist," another Ministry official reportedly said. The former Soviet states need a modern nuclear-security system.[28]

Such a system would have three basic elements:

- *Physical protection:* barriers, sensors, and alarms to deter, delay, and defend against both intruders from outside and theft from inside.
- *Material control:* locked vaults for storage of nuclear materials; portal monitors to prevent workers from walking off the site with nuclear material in their pockets; continuous monitoring of nuclear-material storage sites with tamper-proof cameras, seals, and alarms; and prohibition of access to sites of sensitive materials unless personnel enter in pairs (the two-man rule).
- *Material accounting:* a regularly updated measured inventory of weapons-usable nuclear material, based on regular measurements of material arriving, leaving, lost to waste, and within the facility, including a program to ensure the accuracy of the measurement equipment.

These three elements together are referred to as material protection, control, and accounting (MPC&A). Other desirable elements of a nuclear security system are measures to ensure the reliability of personnel (background checks, training, and dependable salaries for nuclear custodians) and regulation and inspection by an outside agency with real powers of enforcement.

At many former Soviet nuclear facilities none of the three elements of an MPC&A system is adequate. In Russia, according to the Ministry of Internal Affairs, 80 percent of these facilities have no monitors to detect nuclear materials carried through the gates. The army is transporting and storing nuclear materials unsafely, a Ministry of

Atomic Energy official has charged. Some facilities store hundreds or even thousands of kilograms of bomb-grade materials in rooms secured with simple padlocks. Many have broken windows and broken fences. "Even when there are no holes in the fences, protection by the Interior Ministry's Internal Troops is being replaced at a number of facilities ... by old ladies employed by the paramilitary guard detachment because it is believed that it is cheaper." Some are guarded, as one expert put it, by "Aunt Masha with a cucumber." Control and accounting are also weak.[29]

None of the other former Soviet states has nuclear weapons, at least as far as is known. But facilities that store nuclear materials are located in Armenia, Belarus, Georgia, Kazakhstan, Latvia, Lithuania, Ukraine, and Uzbekistan. Especially worrying is a desalinization plant at Aktau, Kazakhstan, across the Caspian Sea from Iran. The power reactor burns 20–25 percent enriched uranium fuel or mixed oxide fuel containing 23.19 percent plutonium. It is capable of producing 110 kilograms of plutonium annually. The site stores "ivory grade" plutonium, which is particularly useful for weapons production because of its low concentration of undesirable isotopes.[30]

The government of Georgia discovered some HEU and spent fuel stored at an obsolete nuclear reactor outside Tbilisi, and requested U.S. assistance in disposing of it. Eventually the U.S. government transported the material to the United Kingdom, where it will be turned into fuel. Another nuclear facility is located in the republic of Abkhazia, which is no longer under Georgian control. U.S. government officials know little about the plant or about how much nuclear material might be stored there.[31]

The principal purpose of the Soviet security system was to keep out American spies. Civilian research facilities, even those which process or store weapons-grade materials, were not considered strategically important targets for potential spies, and have only minimal security.

One staff member of President Yeltsin's Security Council confirmed in interviews what other Russian officials have said in the past: to ensure their ability to meet production quotas under the Soviet system, nuclear facilities often produced extra plutonium to have on hand in case of a future shortfall. As much as 10 percent of production may have been diverted and not entered into the accounting system.[32] This practice of excess production was not considered dangerous from the standpoint of theft because there was no market in Russia for HEU or plutonium. Now, however, there is a growing perception of a lucrative market for nuclear materials. These secret caches of material, likely to be found at many production sites, present a real danger in the current economic climate.

Even the closed cities are experiencing nuclear crimes. Stories in the Russian press allege that depleted uranium, which cannot be used in nuclear weapons but could be mistaken for weapons-usable material, was stolen from Arzamas-16. Local governments in the closed cities have expressed concern about inadequate security for nuclear materials.[33]

Nuclear Theft

Smuggling of nuclear materials is not as serious a problem as many journalists would have us believe, but it is far worse than many Russian government officials are willing to admit. The technical distinction between weapons-grade and weapons-usable nuclear materials has been an important issue in discussions between the United States and Russia about the problem. Russian officials have repeatedly denied that any smuggling case has involved weapons-grade nuclear material, which, according to the strict definition, is uranium enriched to more than 90 percent U-235 or plutonium with less than 7 percent Pu-240. But at least eight cases have involved materials that are weapons-usable, albeit with a less efficient yield than weapons-grade material.

Most reports about smuggling nuclear weapons or weapons-usable components have been unreliable. This is partly a function of journalists' confusion about various distinctions among kinds of radioactive and nuclear materials. One of these is the distinction between weapons-grade material and weapons-usable material. Another is that between nonfissile radioactive isotopes and fissile material used to make nuclear weapons. The former include medical isotopes that, although very toxic, cannot be used to create a nuclear detonation in a bomb (see Chapter 2).

While no single known case has involved enough material to make a bomb, quantities of stolen weapons-usable (but not weapons-grade) HEU have been seized both inside and outside the Russian Federation. Examples include:

- One and one-half kilograms of 90 percent enriched HEU stolen from the Luch production facility at Podolsk in October 1992.
- Nearly two kilograms of 36 percent enriched HEU stolen from a naval base in Adreeva Guba in July 1993.
- Four and one-half kilograms of 20 percent enriched HEU naval fuel stolen from the Murmansk shipyard in late 1993.
- Three separate caches of weapons-usable HEU and plutonium, ranging in size from less than a gram to 350 grams, seized in Germany in the summer of 1994.
- Nearly 3 kilograms of 87.7 percent enriched HEU seized in Prague in December 1994.
- Seven kilograms of HEU reportedly stolen from the Pacific Fleet at Sovietskaya Gavan in January 1996.

Western officials believe that some of the materials seized abroad may also have come from Russia, although Russian officials deny this.[34]

A few cases have appeared to link buyers with sellers. Reports began to surface shortly after the dissolution of the Soviet Union that Iran had purchased components of nuclear weapons, and even intact warheads, from Kazakhstan. The U.S. government looked into the reports and concluded that they had no basis in fact. Subsequently it was reported by a number of U.S. government officials, including Secretary of State Warren Christopher, that Iran had approached Kazakhstan in connection with enriched uranium at the Ulbinsky (Ulba) Metallurgy Plant, a facility for the production of nuclear reactor fuel in northeastern Kazakhstan. However, the accuracy of these reports has been the subject of some dispute.[35] U.S. government experts, in confidential interviews, have recounted official Kazakhstani claims that

Iran approached the Ulba plant about a possible purchase of low-enriched uranium (LEU) but *not* HEU.

In another reported case, Turkish police apprehended a Turkish professor of public policy in the act of selling two and a half kilograms of uranium of uncertain enrichment to three Iranians, said to be agents for Sawama, the Iranian Secret Service. The police claimed the uranium had been brought to Turkey by visiting Russians.[36]

Czech press reports on the trial concerning the December 1994 seizure of HEU in Prague revealed that, according to Czech police, the uranium had been obtained by Russian organized criminals and was bound for a country that was trying to develop nuclear weapons. Czech papers said the investigation had not yet confirmed that the radioactive cargo was bound for Iraq, as had been claimed by the Italian newspaper *Corriere della Sera.*

Konrad Porzner, the head of Germany's Federal Intelligence Service, told a German parliamentary committee in January 1996 that of the thirty-two cases in which interested buyers had been registered by German intelligence in 1995, sixteen involved states. Moreover, Porzner testified that he had definitive proof that Iran and Iraq had been seeking nuclear materials on the black market. The Iranian government has denied the charge, and Porzner refused to answer subsequent questions.[37]

There is clear-cut evidence that fissile materials have been stolen from former Soviet facilities, but there is no evidence (so far) that nuclear weapons have been stolen or that stolen fissile materials have been sold to other nations or to terrorists. Nevertheless, the risk of such transactions is real and troubling.

Chemical, Biological, and Radiological Security

There are many cases of theft of medical isotopes and other sources of radiation. These incidents are often overlooked because radioisotopes cannot be used to make detonable nuclear bombs. But terrorists could use them to draw attention to their cause, to wreak havoc, and to terrorize civilians.

As part of their battle for independence from Russia, Chechens have repeatedly threatened to attack nuclear power plants in Russia. The Russian government formed an interagency group to address concerns about nuclear terrorism, including attacks against nuclear power plants. The group ordered that additional safeguards be put in place at power plants, for example, but the effort was not comprehensive, according to a Ministry of Atomic Energy official. An intelligent terrorist could easily circumvent the new controls, he said.[38]

Inadequate security measures for Russia's chemical weapons stockpile are also worrying. Russia has admitted possessing 40,000 tons of chemical warfare agents, although the amount may be significantly larger.[39] The stockpile includes mustard (a heavy, viscous poisonous agent that was responsible for many deaths and injuries during World War I) and other persistent blistering agents, but most of it is composed of nerve agents, chemicals so toxic that one drop is enough to kill a human being.

Vil Mirzayanov, who worked for twenty-six years at the State Research Institute of Organic Chemistry and Technology, the principal institution responsible for developing chemical weapons in Russia, testified before the U.S. Senate that the environment at Russian chemical weapons facilities is "conducive to foul play." "Warheads with chemical weapons are kept separately from the powder charges, so a potential criminal would not have to worry about accidental explosion." Customs agencies have no equipment for detecting poisonous agents, so "there are very few obstacles to prevent illegal export of chemical weapons from Russia." Russian officials have admitted that they do not know how much chemical agent has been produced.[40]

Perhaps even more troubling is the possibility that senior Russian officials themselves might become involved in illegal exports of chemical weapons or their precursors. Such high-level smuggling may already have occurred. Suspicion has fallen, for example, on Lieutenant-General Anatoly Kuntsevich, who commanded Russia's Chemical Corps and directed a secret Ministry of Defense research installation. The facility was developing a new nerve agent that would be harder to detect with existing sensors. He won a Hero of Soviet Labor medal for his efforts in 1981. After the collapse of the Soviet Union, President Yeltsin named Kuntsevich his advisor for chemical and biological disarmament, a move described by an intelligence analyst as akin to "putting the fox in charge of the chicken coop." Kuntsevich is alleged to have exported 800 kilograms of a reagent used in producing chemical weapons to Syria. The reagent, dichloranhydride methyl phosphonic acid, is used to manufacture a very few rarely produced pesticides, but it is a close chemical precursor of the deadly nerve agents sarin and soman. A further five tonnes was allegedly scheduled to follow. The chemicals were reportedly stolen from military installations in Russia. Kuntsevich has denied any wrongdoing, but one former colleague insists he was guilty of something "the scale of which made the Sarin gas attack on the Tokyo underground last year 'seem trivial.' "[41]

Kuntsevich was one of many suspected criminals who ran for Parliament in 1995 in the hope of gaining parliamentary immunity. He ran on the Liberal Democratic ticket (Zhirinovsky's party), but was removed from the ballot when the seriousness of the charges against him was disclosed.[42] Kuntsevich was eventually acquitted, an outcome that some Russians attribute to his friends in high places.

President Yeltsin confirmed in May 1992 that the Soviet and later the Russian government had been engaged in the illegal development of biological agents. A network of nominally civilian secret biological weapons facilities called Biopreparat employed 25,000 people at eighteen or more R&D sites. Biopreparat specialized in growing agents in antibiotics to make them resistant and thereby ensure that victims could not be cured; its products included a virulent form of encephalitis and a strain of bubonic plague resistant to twenty-six antibiotics. The program to develop a dry form of genetically engineered plague, resistant to antibiotics, was a top priority for Biopreparat during the second half of the 1980s. A single bomb filled with this dried "superplague," dropped on a city of 100,000 persons, would kill half the population, according to Vladimir Pasechnik, a senior Biopreparat microbiologist who defected to the West in 1989. Pasechnik also claims that Biopreparat officials considered using terrorists to deliver the agent: "Terrorists might introduce it and then deny it." Yeltsin promised to shut down the entire illegal program, but international inspectors say he has been unable to keep his promise. Russian scientists reported in 1998 that they had developed a new, genetically engineered strain of anthrax, but it was unclear whether they had done so for offensive or defensive purposes. Existing vaccines may not be effective against this new strain. CIA analysts are reportedly convinced that Russian expertise in biological weapons has already been exported to Iran.[43]

The Chemical Weapons Convention, which entered into force in 1997, requires that all parties destroy their chemical stockpiles within ten years. But the destruction of chemical weapons, it turns out, is very expensive: Russia estimates that destroying its chemical arsenal will cost at least $5.4 billion. Russian government officials repeatedly announce that Russia cannot afford to destroy its chemical stocks. Moreover, members of Russia's budding environmental movement are deeply concerned about the possible release of toxic agents during the destruction process and have opposed the construction of facilities for destroying chemical weapons.[44]

Insecure Borders

The breakup of the Soviet Union created fifteen new countries requiring export-control legislation and fifteen new sets of borders requiring customs police and border guards. Organized crime is taking advantage of permeable borders between Russia and the Baltic states to smuggle stolen cars, weapons, and metals, including radioactive materials. The borders of the southern tier—including Turkmenistan, Kazakhstan, Southern Russia, and Azerbaijan—are completely unguarded in some areas, including points of entry into Iran. An official involved in the U.S. program to upgrade customs and border-control procedures in the region told me: "The borders of the southern tier are so long and so porous. And the bureaucracies are new—without any real spirit. There is no tradition of an honest bureaucracy in these countries. Border guards are not adequately paid—so it's almost impossible to stop corruption. We can and will deal with those countries—but we're not optimistic."[45]

There is regular ferry traffic across the Caspian Sea between Iran and Dagestan in southern Russia, Kazakhstan, Azerbaijan, and Turkmenistan. There is almost no border control on the Caspian, an issue that the U.S. government is "glossing over," according to the official.

The borders between Russia and the rest of the southern tier are also badly guarded—so that materials smuggled out of Russia into Kazakhstan, for example, could easily find their way to Iran. The U.S. government supplied Kazakhstan with three boats that were intended to be used for monitoring ferry and boat traffic, but the Kazakh Ministry of Defense took charge of the boats, and there is still no real policing of Caspian ports.[46]

In February 1995 a team of U.S. Customs officials was invited to visit border crossings in the three Baltic states: Estonia, Latvia, and Lithuania. They found widespread corruption, porous borders, and "virtually no export controls." Trucks were often forced to wait in line for days at points of entry. Drivers set up tables on the side of the road and spent their days drinking and playing cards. At one site in Latvia there was no place to perform even a cursory examination of the cargo or the vehicles. There were no tools, not even a flashlight. "The road was covered with mud and slop . . . The border officers avoided wading through the mud by forcing the drivers to come to them. Obviously no Customs inspector was going to do more than examine the paperwork . . . An apocryphal comment was made about there being only two pairs of felt boots for the Customs officers to wear in the coldest weather. They had to pass them from shift to shift and finally wore them out, thus ending any further cold weather inspections."[47]

Chaos, Crime, and Corruption

Russia is undergoing a crisis of statehood—a faltering metamorphosis from totalitarian empire to law-based, federal state. Many nations, including the United States, have been through similar transitions. As a specialist on Russian law has put it: "The United States' transition from colony to sovereign republic . . . was long, difficult and eventful . . . Along the way there was dissent, repression, revolution, [and] war . . . Russia's journey too is likely to be long, arduous, and painful, and the outcome may not be clear for quite some time."[48] What is unique about Russia and the rest of the former Soviet Union is that this fragmenting empire possesses tens of thousands of nuclear warheads, hundreds of tons of fissile materials, and tons of toxic agents that may, if not carefully guarded at all times, end up in the hands of terrorists or of hostile states.

Chaos (*khaos* in Russian) has menaced the Russian state since its inception. The perception that chaos lurks beneath the surface of the sociopolitical order has been more enduring than any sociopolitical order itself. With the end of Communism, chaos and lawlessness have reasserted themselves in full force. The territorial integrity of Russia is threatened, not by an external enemy, but by Chechnya, an internal, fractious Russian region bent on secession. Organized crime and corruption have penetrated the highest levels of government. The Russian people are thoroughly disgusted with the government's privatization campaign, which was accomplished in part via closed auction, transferring much of Russia's productive capacity from "the people" to the chosen few.[49]

Suspected criminals brazenly ran for Parliament, knowing they would be granted parliamentary immunity if they won. Duma members insist on carrying their guns into the parliamentary building, where there have been several widely publicized shootouts. They are also in the habit of passing laws that violate the constitution, says a critic in the upper house of the parliament.[50] Journalists are shot at an alarming rate—especially those investigating corruption. According to the International Federation of Journalists, more than half of the sixty-one journalists killed in the world in 1995 were killed in Russia.[51]

Organized crime has infiltrated law enforcement agencies, commercial banking, and the political and military leadership, especially outside urban centers. Organized crime controls more than 40,000 businesses, including 500 joint ventures and 550 banks, reports a financial newspaper based in Russia. The "shadow" economy makes up a significant fraction of Russia's GDP, suggesting that the economy is in far better shape than official statistics would suggest, but that it is highly criminalized. And analysts believe that the Mafia controls nearly half of Russia's consumer markets, real estate, and banking sectors.[52]

Russian criminal groups are in a class of their own, say U.S. law enforcement personnel. They have unparalleled access to computer technology, encryption techniques, and money-laundering facilities capable of processing hundreds of millions of dollars. Their members include Ph.D. scientists and former senior KGB agents with access to sophisticated weapons. Networks between military units and organized crime are facilitating the export of conventional military hardware, some of it to drug traffickers and terrorists. Russian criminals have attempted to sell Colombian drug traffickers a submarine, helicopters, and surface-to-air missiles. Another criminal group, whose members include high-level Russian officials, is exporting weapons to Kurdish terrorists through Poland, Bulgaria, Slovakia, and the former Yugoslavia, according to the CIA. These same networks might be used to export, in a much more efficient fashion than has been seen so far, fissile materials or precursors of chemical weapons, or even the weapons themselves.[53]

Russian organized-crime groups "hold the uniquely dangerous opportunity to procure and traffic in nuclear materials." Some analysts believe they are already involved in exporting weapons of mass destruction. Iran is alleged to have worked through an unnamed Russian criminal group to hire Russian scientists with expertise in nuclear, chemical, and biological weapons, and to have obtained raw materials for producing these weapons from the same group.[54] Reports surface from time to time implicating organized crime in nuclear smuggling, although most of these cases have turned out to involve nonfissile radioactive materials. If organized crime and corrupt government officials are in fact involved in smuggling WMD, the challenge for governments is to make trafficking in WMD unprofitable. This will require strengthening international cooperation in the areas of law enforcement and intelligence.

Russia is simultaneously buffeted by both centrifugal and centripetal forces. Russians are understandably confused about where their country begins and ends: a Russian state has never before existed within the current borders. Late imperial Russia and the Soviet Union both encompassed a significantly larger territory. The Eastern Slavic states of Ukraine and Belarus are arguably more Russian than much of what is now the Russian Federation, and ethnic Russians consider themselves closer psychologically and culturally (as well as linguistically and geographically) even to the Western Slavs (Czechs and Poles) and the Southern Slavs (Serbs, Croats, and Bulgarians) than to the aboriginal peoples of Siberia and the Caucasus. Chechnya is now the most notorious of the restive Russian regions, its bid for independence having led to a protracted civil war.

The Russian government too seems puzzled about where Russia begins and ends—both psychologically and legally. Of the fifteen new countries formed from the former Soviet Union, Russians are guarding the borders of nine, creating confusion about who is responsible for export controls and customs. There are approximately 5,000 Russian border guards stationed in Georgia, 5,000 in Armenia, 100 in Kazakhstan, at least 150 in Turkmenistan, 3,000 in Kyrgystan, 8,000 in Tajikistan, 50 in Uzbekistan, and an unknown number in Belarus. Belarus, Kazakhstan, and Russia signed a customs union in 1995, agreeing that no troops would guard their mutual borders. (Russia later announced that it would maintain some customs guards on the border with Kazakhstan.) While the agreement is a boon for trade, it may make it difficult to stop the flow of illicit exports, including nuclear materials or precursors to chemical weapons.[55]

The thefts of nuclear materials have made clear that nuclear security is inadequate in the former Soviet states. But the thefts that have come to the attention of authorities so far have been committed by amateurs—plant employees and petty criminals with no idea how to market the material they steal.

A far more serious danger is posed by the prospect of corrupt government officials trading in nuclear, chemical, or biological weapons, components, or expertise. Given the practice of hiding nuclear materials to be sure of meeting production quotas, there are probably secret caches of nuclear material not yet known to the West, perhaps known only to a few officials. Corrupt government officials working with organized criminals know how to market goods and services. And unlike those involved in nuclear smuggling so far, they also know whom to bribe to avoid being caught. Organized criminals are already selling conventional weapons to terrorists and drug traffickers. Those same networks could be used to trade in weapons of mass destruction.

The State as Terrorist

> My sense used to be that biological weapons were rather exotic . . . I didn't really think of it as a weapon that anyone would prepare to use. Iraq may show—and give inspiration to others—that this is a viable weapon, it is a weapon worth having.
>
> —Ambassador Rolf Ekeus, 1997

Within a week of the cease-fire that ended the Iran-Iraq War on August 20, 1988, Saddam Hussein unleashed his arsenal of chemical weapons against the Kurds, an ethnic minority in Northern Iraq that had long resisted domination by Baghdad.[1] Saddam used both conventional and chemical weapons to break the Kurdish insurgency and force the rebels and thousands of civilians to flee to Turkey. One eyewitness said: "The planes dropped bombs. They did not produce a big noise. A yellowish cloud was created and there was a smell of rotten parsley or onions. There were no wounds. People would breathe the smoke, then fall down and blood would come from their mouths." Another witness reported: "In our village, 200 to 300 people died. All the animals and birds died. All the trees dried up. It smelled like something burned. The whole world turned yellow." Thousands of civilians died. No sanctions were imposed on Iraq by either the United States or the United Nations for its repeated use of chemical weapons, first against Iran and then against its own citizens.[2]

By the time these acts of chemical terrorism were committed, the international community was well aware of Iraq's sizable supplies of chemical weapons, but concern that Iran might win the war stifled international response. Acquisition of chemical weapons was permitted at the time under international law, but use was banned by the Geneva Protocol, which Iraq had signed and ratified. Attacking noncombatants was also prohibited.[3]

In retrospect it is clear that the international community made a grave error in ignoring Iraq's acts of chemical terrorism. Saddam continued his weapons programs relatively unimpeded. Lack of reprisal for his early transgressions seems to have contributed to his impression that future violations would also be ignored. Saddam seemed genuinely surprised by the international Coalition's response to his invasion of Kuwait. The indifference of the world community to Iraq's repeated violations of international law could be interpreted by other states as a green light to use weapons of mass destruction (WMD), as long as the weapons are used only in the developing world.[4]

Besides having used chemical weapons in acts of state terror against its own citizens, Iraq is known to sponsor terrorism outside its borders. Despite its military defeat in the Gulf War and its commitment to destroy its WMD, Iraq has the potential to use these weapons again in acts of terrorism around the world. It has reportedly made threats to smuggle anthrax and other WMD into Britain, in one case threatening to put anthrax in duty-free alcohol, cosmetics, cigarette lighters, and perfumes.[5] Iraq has become the template for rogue states because of its belligerent behavior, its transgressions of international norms, its violations of human rights, and its attempts to provoke the United States. The case of Iraq also shows how little can be done to deal with the problem of the proliferation of WMD. Preventive war did little to eradicate Iraq's arsenal, and the most intrusive inspection regime ever devised has left inspectors guessing, especially about biological weapons. Saddam has demonstrated the havoc that can be wreaked by WMD and the impotence of the international community in responding to their use.

Saddam is continuing his pursuit of WMD despite punishing sanctions imposed by the United Nations that cost Iraq billions of dollars in forgone export earnings. Among the biological agents stockpiled by Iraq is aflatoxin, which is useful only for terrorism: it is a slow-acting carcinogen with no utility as a weapon on the battlefield.[6] Officials can only speculate about how Iraq might intend to use such an agent—perhaps against its own ethnic minorities, perhaps against Israel, perhaps in terrorist acts in the United States or Europe. Iraq reportedly told the United Nations it had made enough botulinum toxin to wipe out the earth's population several times over. The CIA warned before the Gulf War that Saddam had the ability to deliver biological or chemical agents clandestinely, using special forces or foreign terrorists, to "take out as many of his enemies with him as he could" if he felt sufficiently threatened. The CIA's warning still applies, despite concerted efforts to destroy Iraq's weapons of mass destruction.[7]

Proliferation and Terrorism

In 1981 Kenneth Waltz presented an elegant argument in favor of the proliferation of nuclear weapons. Nuclear weapons induce caution, Waltz argued, thereby reducing the likelihood of war. The probable cost of attacking an adversary that possesses nuclear weapons makes war impossible to contemplate. Thus nuclear proliferation promotes peace. A number of other "proliferation optimists" have agreed with Waltz that nuclear weapons prevent war. If all nations had nuclear arms, according to one study, the probability of nuclear conflict would be reduced essentially to zero, provided all acquired their nuclear weapons at approximately the same time.[8]

Scott Sagan, a "conditional proliferation pessimist," presents a dissenting view. He argues that nuclear proliferation is dangerous because professional military organizations may not pursue rational policies of deterrence. New nuclear powers are more likely to fight preventive wars, and are themselves vulnerable to preventive strikes, especially during the period when their nuclear weapons capabilities seem imminent. They are more likely to build vulnerable second-strike forces. And they are less likely to invest in adequate command and control mechanisms, so their arsenals may be prone to accident or unauthorized use.[9]

None of these analysts has pointed out the most important danger of proliferation from the point of view of terrorism. New nuclear states with small arsenals are more likely to use these weapons against noncombatants—in other words, in terrorist mode—whether or not they are "rogue states." Targeting noncombatants is, quite simply, the most efficient use of nuclear weapons. Recall General Marshall's explanation of why the United States used atomic bombs against Hiroshima and Nagasaki. Some planners wanted to drop the bomb in the Sea of Japan, Marshall claimed. "Others wanted to drop it in a rice paddy to save the lives of the Japanese." But the United States had only two nuclear weapons in its arsenal, and the most efficient (and most shocking) way to use them was to drop them on civilians in cities.[10] Chemical and biological weapons, too, are most efficiently used against noncombatants, since armies can be protected against their lethal effects.

Proliferation and the threat of terrorism using weapons of mass destruction are intimately linked. Iraq, North Korea, Syria, Iran, and Libya, all listed by the U.S. State Department as sponsors of terrorism, are also all suspected of pursuing WMD. North Korea, for example, has been exporting equipment that, while ostensibly for benign purposes, could be used to manufacture WMD. It is not inconceivable that North Korea may eventually find it profitable to sell such equipment to terrorists. Moreover, continuing proliferation of WMD is likely to erode norms against the use of these weapons, including use for terrorist purposes. Conversely, efforts to prevent further proliferation could have a salutary effect on efforts to fight WMD terrorism. The smaller the world's arsenal of WMD, the harder it will be for terrorists and their state supporters to acquire them. And domestic legislation required under the bans of chemical and biological weapons makes it easier for law enforcement agencies to prosecute terrorists. It is therefore critical that parties to these treaties adopt that required legislation.

Nonproliferation Agreements

Arms control treaties enhance security in two ways. They strengthen norms against the weapons that are banned, and they increase confidence that adversaries will not acquire the weapons. In the 1960s, when France first got nuclear weapons, these weapons were seen as symbols of state power and prestige. Over time the Nuclear Nonproliferation Treaty (NPT), which entered into force in 1970, has contributed to changing that image. Had Belarus, Kazakhstan, and Ukraine kept the nuclear weapons they inherited with the breakup of the Soviet Union, the weapons would have detracted from their image; they would have joined the ranks of pariah nations like Iraq and North Korea.[11] The long-term impact of India's and Pakistan's decisions to test nuclear weapons in 1998 is not yet known.

Arms control treaties enhance international security without worsening the security dilemma: they make the dilemma of security less vicious. Each state would feel most secure if it alone had WMD while its adversaries had none. But this situation would be unlikely to persist for long; proliferation tends to induce further proliferation. Second best is for neither adversary to have WMD. Arms control treaties that include provisions for inspection help reassure states that they will receive early warning if their adversaries are acquiring the weapons.[12] Even treaties with apparently comprehensive inspection regimes, however, provide no guarantees. Iraq has successfully exploited weaknesses in several treaties that ban WMD.

The Nuclear Nonproliferation Treaty

In 1963 President Kennedy predicted that there could be ten nuclear powers by 1970 and perhaps fifteen to twenty by 1975. In 1998, however, only eight states were believed to possess nuclear weapons. The NPT is at least partly responsible for inhibiting the spread of nuclear weapons.[13]

Under the NPT, only the United States, the United Kingdom, France, the Soviet Union, and China are allowed to possess nuclear weapons, while the other parties to the treaty agree not to develop them in return for access to nuclear power technology and the promise that the five nuclear states will reduce and eventually eliminate their nuclear arsenals. Most of the industrialized countries also adopted nuclear-export regulations and placed nuclear exports under International Atomic Energy Agency (IAEA) safeguards.

The IAEA is charged with monitoring compliance with the NPT. The IAEA has two assignments. The first is to provide technical assistance to promote the peaceful use of nuclear energy. The second is to monitor the activities of nations that do not have nuclear weapons to ensure that nuclear-energy technology is not diverted to weapons programs. As David Kay has noted, giving the IAEA two roles was a bad idea from the start. In the case of Iraq, Kay says, the IAEA's technical assistance "went directly to individuals and activities later identified with the clandestine programs."[14]

Iraq exploited several weaknesses in IAEA safeguards. Traditionally, IAEA inspectors behaved more like accountants than detectives. Iraq was carrying out prohibited activities at sites regularly visited by IAEA inspectors, such as Al Tuwaitha, and also at a network of secret nuclear facilities. Although the 1972 model NPT safeguards agreement provides for special inspections at undeclared sites (sites not subject to routine inspections under the NPT), before the revelations about Iraq's nuclear program such inspections had never been carried out. They were not part of the IAEA's modus operandi. Dr. Jaffar Dhia Jaffar, a British-trained physicist and the head of Iraq's nuclear weapons program, boasted to IAEA inspectors that he could give a seminar on "How to beat the NPT."[15]

Another weakness of IAEA safeguards is that at the time they were concerned principally with materials required for the manufacture of nuclear weapons, rather than with facilities capable of developing weapons.[16] Parties were required to provide information about the design of nuclear facilities "as early as possible," before nuclear materials were present at the facilities. In practice, parties provided design information 180 or fewer days before nuclear material had been moved to a site. North Korea notified the IAEA seven years *after* nuclear material was in place. According to one former inspector, "practices like this opened the way for Iraq to claim that it should not be faulted for not declaring its centrifuge enrichment plant since no material had been introduced and it would have transmitted design information in due course."[17]

The Biological Weapons Convention

Biological agents are significantly easier to produce than nuclear weapons. The equipment necessary for manufacturing biological agents also has civilian uses, and is thus widely available from commercial sources.

(This applies equally to chemical weapons.) Iraq has been able to acquire equipment from many international sources, including agricultural sprayers from Italy that could be used to circumvent some of the problems of disseminating biological organisms in respirable aerosols.

Advances in biotechnology have made it easier to manufacture and disseminate biological agents. For example, scientists now have a more comprehensive understanding of the optimal growth requirements of organisms—temperature, nutrient concentration, pH, and so on—and since the early 1980s fermenters have been designed to meet those requirements. Not surprisingly, proliferation of these weapons has increased. But nowhere outside Russia is a biological weapons program as advanced as Iraq's.

The 1972 Biological Weapons Convention (BWC) bans biological weapons outright. Unlike the NPT, it does not divide nations into "haves" and "have nots," so the BWC is potentially more attractive to developing countries. But it has no provisions for inspection or enforcement, and there have been a number of violations, most notably in Russia and Iraq. The Ad Hoc Group of BWC member states has been considering adding inspection provisions to this treaty. These provisions would make proliferation more risky and more expensive, although they would not necessarily uncover all cheating.

The Chemical Weapons Convention

Until the mid-1980s the only restriction on chemical weapons was the 1925 Geneva Protocol, which prohibited their use. It did not prohibit the production, stockpiling, or transfer of weapons or technology. Thus no international law forbade companies in Germany and elsewhere to assist Egypt, Iran, Iraq, Libya, and Syria with the production of chemical agents. Moreover, the protocol did not provide for verification of compliance or for sanctions in the case of violations. And many nations, including Iraq, reserved the right to retaliate in kind if another country, whether a signatory to the treaty or not, used chemical weapons first. Thus the protocol was effectively a no-first-use agreement.

In 1984, prodded by Iraq's use of chemical weapons against Iran, western industrialized nations formed the Australia Group to harmonize national export controls on the material and equipment used in chemical and biological weapons and to exchange information about proliferation. The Australia Group has not prevented proliferation of chemical weapons, but may have slowed it down. Before its enhanced export controls were implemented, Iraq was able to field nerve agents just a year after its 1983 deployment of mustard gas. In comparison, Iran, which first used mustard gas in 1987, has reportedly still not developed nerve agents.[18]

The Chemical Weapons Convention (CWC), which bans the development, production, acquisition, storage, transfer, and use of chemical weapons, entered into force on April 29, 1997. This convention has the most intrusive verification measures ever devised. Thousands of chemical manufacturers will have to open their doors to international inspectors on a routine basis. Governments, even those in compliance, may be subjected to on-site, short-notice inspections at facilities so sensitive that their very existence has not been publicly revealed. Parties are required to control exports of some commercially important chemicals to nonparties. And they are required to enact legislation to penalize citizens who violate those controls.

Unlike the Geneva Protocol, the CWC bans all use of chemical weapons, not just first use. This makes it impossible for states to justify using chemical weapons by accusing their enemies of having used them first, as Iraq falsely accused Iran during the Iran-Iraq War.[19] And as with the Biological Weapons Convention, there are no "haves" and "have nots": all parties must comply with every one of the treaty's provisions.

Coupled with the Australia Group, the CWC has good prospects for inhibiting the future proliferation of chemical weapons. Several nations that have or are suspected of having developed chemical weapons, such as India, Pakistan, South Korea, South Africa, China, and Iran, have ratified the CWC. These countries will have to open up their previously secret programs to international inspectors and destroy their stockpiles. It remains to be seen how useful the treaty will be in rolling back the estimated twenty to twenty-five chemical weapons programs that are active around the world. As the case of Iraq makes clear, it is far easier to detect production of chemical agents than to ferret out chemical weapons that have already been produced.

Iraq's WMD Programs

Nuclear Weapons

Iraq signed the NPT in 1968, the same year it began operating a small, Soviet-supplied research reactor at the Al Tuwaitha Nuclear Center near Baghdad. Between 1974 and 1976 Iraq negotiated nuclear cooperation agreements with France and Italy, and soon afterward it bought two reactors from France and a set of reprocessing laboratories from Italy. Both reactors, a large-megawatt research reactor and a critical assembly, were fueled by HEU supplied by France. If diverted for weapons use, the fuel for these reactors could have produced several nuclear weapons. In the early 1980s Iraq also purchased several hundred tons of natural uranium from Portugal, Niger, Brazil, and Italy, which could be used as the raw material for the uranium specimens that would be irradiated in the reactors. These purchases would have allowed Iraq to reprocess plutonium for use in nuclear weapons, and to use the HEU fuel as a short-term expedient.[20]

Israel, alarmed at the scale of Iraq's nuclear purchases, began a concerted effort in 1978 to halt these transactions. When diplomacy proved futile, Israel's foreign intelligence service, the Mossad, began a campaign of sabotage, assassination, and psychological warfare to stop the European firms from fulfilling their contracts. In April 1979 reactor cores bound for Iraq were damaged by saboteurs; in June 1980 employees of the Iraqi Atomic Energy Commission were murdered in Paris; in August 1980 French and Italian nuclear firms were attacked and their employees threatened. After the failure of diplomacy and covert operations to derail Iraq's nuclear weapons program, Prime Minister Menachem Begin decided to bomb the large reactor that Iraq had purchased from France. On June 7, 1981, only months before the reactor was due to become operational, Israeli F-16s destroyed it in a surprise raid. This attack was a stunning blow to Saddam Hussein's dreams of nuclear weapons. Undaunted, however, he ordered a new weapons program incorporating lessons learned from this experience.[21]

Iraq's new approach to the production of nuclear weapons was to develop its own ability to produce HEU.[22] In 1987 construction began on two identical facilities at Al Tarmiya and Ash Sharqat, which were to house systems called calutrons, which use electromagnetic radiation to separate uranium isotopes. Electromagnetic isotope-separation technology was developed in the 1940s during the Manhattan Project. The United States abandoned calutrons in favor of a more efficient method, gaseous diffusion, and the calutron technology was declassified. Iraqi scientists created their own version of the calutron, the "Baghdadtron." Each of the two plants was designed to produce about fifteen kilograms of HEU a year, but neither was completed by the time of the Gulf War. Iraq's electromagnetic isotope-separation program operated undetected for several years before the Gulf War. The calutrons, when operational, would have allowed Iraq to produce significant quantities of weapons-grade uranium fairly rapidly.[23]

Iraq was simultaneously pursuing centrifuge technology, another technique for enriching uranium. In 1989, with extensive foreign assistance, it began constructing a centrifuge factory at Al Furat. The Iraqis claimed the factory could have produced 200 centrifuges a year by 1992 (if the Coalition's bombing had not put an end to production), but IAEA inspectors believe the rate could have been as high as 2,000. By 1991 Iraq had a workable centrifuge design and "was making good progress towards the establishment of a mass production capability," according to the IAEA.[24]

After its invasion of Kuwait Iraq launched a crash program to construct a crude nuclear device using French- and Soviet-supplied HEU fuel, even though that fuel was subject to IAEA inspections. The Iraqis initially claimed that they could not have completed a nuclear device until 1994, but eventually they admitted that a "cold test" of the weapon design without HEU could have occurred in 1991. However, the indigenously produced HEU would not have been available until late 1992.[25]

Iraq claims that all work on this crash program was halted by the onset of the Coalition bombing and that it destroyed the pilot reprocessing plant to hide the program from the IAEA. Iraq may have attempted to reconstitute this program shortly after the end of the

Gulf War but, David Albright believes, was foiled by IAEA inspections.[26]

Chemical Weapons

Iraq claims that its original chemical weapons program began in the mid-1970s, was abandoned in 1978, but was resumed in 1980. Iraq started producing mustard gas in 1981 and nerve gas in 1984. In March 1995 Iraq said it had produced 2,850 tons of mustard gas, 210 tons of tabun, and 790 tons of sarin. It began working on VX in 1985 and has admitted producing 3.9 tons of that agent.[27]

The Al Muthanna State Establishment, known to the outside world by the name of a nearby town, Samarra, was the hub of Iraq's chemical weapons program. Also referred to by Iraqis as the State Establishment for Pesticide Production, it comprised a production complex at Al Muthanna, three intended precursor-production sites at Al Fallujah, and munitions stores and a test site at Al Muhammediyat.[28]

The development of Iraq's chemical weapons program can be broken down into three overlapping phases. The first focused on the deployment of massive numbers of tactical chemical munitions. The United Nations Special Commission (UNSCOM) would eventually find and destroy 28,000 filled chemical munitions, mostly artillery shells, rockets, and bombs. In the second phase Iraq emphasized self-sufficiency in its production of chemical agents and precursors. In the third phase it began extending the shelf life of its chemical agents, producing a strategic stockpile including VX, and developing delivery systems for strategic chemical weapons such as warheads.[29]

Iraq has claimed that the launch authority for chemical- and biological-armed missiles was predelegated in case Baghdad was struck by a nuclear weapon. However, UNSCOM notes, "Certain documentation supports the contention that Iraq was actively planning and had actually deployed its chemical weapons in a pattern corresponding to strategic and offensive use."[30]

As far as is known, Iraq was the first country to use nerve agents on the battlefield. Eight separate U.N. investigations confirmed Iraq's use of chemical weapons in the Iran-Iraq War. Chemical agents, David

Segal concluded, were "quite effective in neutralizing Iranian operations." According to Tom McNaugher, chemical weapons played a role in "sharply lowering the morale of Iranian citizens and soldiers . . . This suggests that the effectiveness of these technologies . . . rested on factors lying beyond their technical capabilities." Charles Duelfer, deputy director of UNSCOM, says that "Iraqis believe that chemical weapons saved their country" in the war with Iran.[31]

Biological Weapons

The Iraqi government claims that it started working on biological weapons in 1974, stopped in 1978, and did not resume work until 1985. However, according to an international panel of experts convened by UNSCOM, Iraq's biological weapons program began in the very early 1970s and "suffered only minor hiccoughs over its 20-year life." The program was revived at the Al Muthanna research site by 1985 and shifted to Al Salman (Salman Pak) in May 1987. In the late 1980s Iraq began producing botulinum, aflatoxin, and anthrax. It has admitted conducting research on trichothecene mycotoxins; circumstantial evidence suggests it has produced large amounts of these toxins. Iraq has also produced *Clostridium perfringens* (gas gangrene) and ricin for use against humans and wheat cover smut for use against crops, and has conducted research on haemorrhagic conjunctivitis virus, rotavirus, and camelpox virus. By the end of 1990 it had field-tested rockets, shells, and bombs with botulinum, ricin, anthrax, and aflatoxin.[32]

After its invasion of Kuwait Iraq engaged in what it called a crash program to produce large amounts of biological weapons. Iraq tested a 2,000-liter anthrax spray tank for mounting on aircraft or remotely piloted vehicles; the test apparently failed, but three more spray tanks were produced and stored (Iraq says they were destroyed). Iraq reportedly tried to equip a MiG-21 jet fighter with spray tanks and make it operable by remote control. The goal, U.N. officials claim, was to have the plane fly over Israeli territory, begin releasing biological agent upon crossing the border, and crash land inside Israel. Baghdad filled 100 aerial bombs with botulinum, 50 with anthrax, and 16 with aflatoxin.

In addition, 13 warheads were filled with botulinum, 10 with anthrax, and 2 with aflatoxin. These weapons were deployed in early January 1991 at four locations. In total, Iraq claims it produced 19,000 liters of botulinum, 8,500 liters of anthrax, 2,200 liters of aflatoxin, and 340 liters of *Clostridium perfringens.* Much of this Iraq had loaded into weapons. It claims to have destroyed all its biological-agent stocks and munitions after the Gulf War.[33]

It is important to note that until 1995 Iraq denied having any offensive biological weapons program. Since then Baghdad has repeatedly changed its declarations, gradually admitting to a frighteningly sophisticated program.

For example, it has produced high-quality liquid slurries of anthrax and botulinum toxin. Agricultural foggers bought from Italy were capable of disseminating these slurries in respirable aerosols. Iraq is also known to have bought drying and milling equipment. Some dry agent was loaded into bombs and missile warheads, and Iraq tested aerosol generators aboard helicopters in 1988. Whether these dry-agent delivery systems would have worked is unclear. William Patrick, who worked on the U.S. program to develop biological weapons before President Nixon shut it down in 1969, believes the techniques Iraq employed for disseminating biological agents were crude and "far from successful."[34]

Ambassador Rolf Ekeus, who was then the executive chairman of UNSCOM, describes Iraq's efforts:

> Iraq admits to having filled twenty-five warheads with agent. What we don't know is whether they had figured out how not to kill the agent when the bomb explodes. Or, in the case of the warheads, whether they had figured out how to make the warhead open up before it disappears into a hole. The missile is coming down at Mach 3—three times the speed of sound. It's not easy to get the warhead to open before impact. They would need a proximity fuse—there are strong doubts that Iraq had such an advanced capability. Or they might have tried retarding the speed of the warheads with parachutes, which would have been easier, but we're not sure they knew how to

> use them, or [how to] open missile warheads remotely. Delivering biological weapons in bombs is much easier than from warheads, but we don't even know whether the Iraqis could manage bombs.[35]

UNSCOM still is not certain of the extent of Iraq's biological weapons program, especially its dissemination capabilities. Given Baghdad's habit of admitting to its least sophisticated efforts first while concealing its most advanced ones, it may be premature to conclude that Iraq has only produced crude weapons with wet agents. Ambassador Richard Butler, who replaced Ekeus as UNSCOM's executive chairman in 1997, admits to knowing disturbingly little about Iraq's biological weapons program.[36]

Radiological Weapons

At the end of 1987 the Al Muthanna State Establishment and the Nuclear Research Center at Al Tuwaitha began exploring the use of radiological "area denial" weapons to prevent enemy forces from entering certain territory.[37] Iraq conducted field experiments of lead-shielded containers weighing 1,400 kilograms, loaded with about a kilogram of irradiated zirconium oxide. Three prototypes were tested in 1987, one in a ground-level static test and two dropped from aircraft. Iraq subsequently modified 100 bombs for use as radiological weapons. These bombs weighed only 400 kilograms so they could fit in bomb bays of aircraft. Iraq claims the tests were unsatisfactory and the program was abandoned in mid-1988. The fate of the 100 bomb casings specially constructed for use as radiological weapons is unknown.

The Gulf War and After

In August 1990 Iraq invaded Kuwait. The U.N. Security Council demanded that Iraq withdraw by January 15, 1991. When Saddam Hussein's troops did not withdraw, a coalition force—principally from the United States, Saudi Arabia, Great Britain, Egypt, Syria, and France—attacked Iraq. The war lasted six weeks. Saddam's forces withdrew from

Kuwait and accepted a permanent cease-fire on April 6, 1991. As part of the cease-fire agreement, Baghdad agreed to pay reparations to Kuwait, to reveal the location of its chemical and biological weapons facilities, and to eliminate its weapons of mass destruction.

The apparent success of the Coalition's aerial campaign against Iraq in the Gulf War seemed to herald what the Pentagon calls a "revolution in military affairs." But in fact Iraq's WMD capabilities largely survived the war, demonstrating the limited effectiveness of precision strikes against WMD programs in the absence of perfect intelligence.

In mid-January 1991 the U.S. target list included only two nuclear facilities: the large complex at Al Tuwaitha and a uranium-ore mine northwest of Baghdad. By the end of the war in February, eight nuclear targets had been identified. Eight months later IAEA inspectors had uncovered a total of twenty-one nuclear facilities in Iraq. Even when WMD production facilities were found, there was no guarantee that they would be destroyed. Of the eight nuclear sites identified during the Gulf War, the Defense Intelligence Agency assessed that only five were destroyed, two damaged, and one still operational. And in many cases Iraq had removed vital equipment from the sites long before the start of hostilities. Al Atheer, a site critical to Iraq's nuclear program, was not identified as a nuclear site until the end of the war and was only lightly damaged. Iraq's centrifuge program was completely untouched by the forty-three days of intense aerial bombardment. Three facilities at which Iraq admitted having produced biological agents survived intact, allowing Baghdad to continue working on biological weapons after the war. Coalition forces successfully bombed two biological weapons facilities, but Saddam had already withdrawn all important equipment from those sites.[38]

Contrary to President Bush's assertion following the war that "our pinpoint attacks have put Saddam Hussein out of the nuclear bomb-building business for a long time to come," the Pentagon's comprehensive analysis of the air war against Iraq, the Gulf War Air Power Survey, concluded that the Coalition air strikes merely "inconvenienced" the Iraqi nuclear weapons program. This somber assessment was one of the factors that led the Pentagon to improve its ability to fight an enemy equipped with nuclear, biological, chemical, and missile weapons.[39]

IAEA and UNSCOM Inspections

In April 1991, after the Gulf War, the United Nations Security Council approved a resolution outlining the terms for the cease-fire with Iraq. Resolution 687 requires Iraq to destroy all of its nuclear, chemical, and biological weapons; all missiles with ranges greater than 150 kilometers; and all associated support, research and development, and production facilities. In addition, Iraq is prohibited from developing such weapons in the future. The United Nations Special Commission (UNSCOM) was created to implement the resolution's provisions. The Security Council gave the IAEA the lead on the nuclear program, however, with assistance provided by UNSCOM. The resolution prohibits Iraq from resuming its oil exports until it complies with the disarmament provisions.[40]

UNSCOM and the IAEA initially believed that Iraq would cooperate with their inspections. This optimism proved to be unfounded. Iraq sought from the beginning to conceal from international inspection as much of its nuclear, chemical, biological, and missile programs as possible.

In late June 1991 David Kay and seventeen other inspectors from the IAEA arrived at the front gate of a military site near Al Fallujah that was suspected of housing uranium-enrichment equipment. Kay and his colleagues were on a mission to uncover Iraq's nuclear weapons program in accordance with Resolution 687, which had been approved only two months earlier. The first IAEA inspectors, who had arrived the previous month, had examined several suspicious sites, but had found no concrete evidence that Iraq had conducted a nuclear weapons program or was hiding nuclear-related equipment prohibited by the U.N. resolution.[41]

Instead of giving Iraq six hours' warning of the inspection, as had been done in the past, the team arrived for the inspection without notice. But an Iraqi military officer refused them entry to the Al Fallujah compound. According to the U.N. resolution, the inspectors were supposed to be granted immediate access to any site. After a prolonged standoff, the officer finally allowed several inspectors to climb to the top of a water tower that stood just outside the facility. From this perch the inspectors spotted dozens of trucks loaded with cargo driving out through the back gate. They radioed down to Kay and the others, who sped off in their vehicles to chase the Iraqi convoy. As the inspectors caught up with the fleeing trucks, Iraqi soldiers fired over their heads. The inspectors were able to photograph giant, twelve-ton magnets lashed to the backs of the trucks. Although the inspectors were stopped and searched, they managed to keep the incriminating film.

These magnets were the heart of Iraq's calutron devices, which the Iraqis intended to use to enrich uranium for use in nuclear weapons. The photographs were the first solid evidence of Iraq's secret nuclear weapons program. The scope of the program, however, was not revealed until September 1991, when IAEA inspectors, again led by David Kay, raided the Nuclear Design Center in downtown Baghdad and recovered four boxes of the program's records. The Iraqis confiscated these materials, and returned only some of them.

Next the inspection team visited the headquarters of Petro Chemical-3, the code name of Iraq's nuclear weapons program. At this site the Iraqis were better prepared. When the inspectors emerged from the building with boxes of incriminating documents, soldiers trapped them in a parking lot and demanded the return of the material. The team's only means of communication with the outside was a satellite telephone, which Kay used to contact his superiors and give interviews to the media. Finally, after eight days, the Iraqis relented and allowed the inspectors to leave the parking lot with more than 50,000 pages of documents on the weapons program. These documents included personnel records, procurement records, and details of Iraq's nuclear research. One of the most important was a top-secret report from the weapons lab at Al Atheer, which showed that Iraq had been working on a design for a nuclear bomb.

Current WMD Capabilities

The IAEA warns that Iraq probably still has a "theoretical capability to produce nuclear-weapons-usable material, to fabricate nuclear weapons, and to design and manufacture a missile delivery system." If sanctions and international inspections were to end, the CIA estimates, Iraq could produce enough fissile material for an atomic bomb in five to seven years. According to Rolf Ekeus, the former head of UNSCOM, if the Iraqis acquired enough HEU, perhaps from the former Soviet Union, they would be able to create a "viable weapon" using their previous experience with nuclear weapons design.[42]

UNSCOM also oversaw the destruction of Iraq's chemical weapons program. Over a two-year period UNSCOM destroyed more than 480,000 liters of chemical agents, more than 28,000 chemical munitions, and nearly 1.8 million liters of precursor chemicals. However, the most technologically advanced elements of the chemical weapons program remain unaccounted for. Iraq may still be concealing stocks of VX, enough precursors to produce 400 tons of the nerve agent, undeclared munitions, and VX-production facilities.[43]

The CIA reports that although chemical facilities were damaged by Coalition bombing, Iraq was able to hide key production equipment, and that if U.N. sanctions were eased Iraq would be able to produce chemical weapons "almost immediately." As many as 100 chemical weapons facilities may remain, according to U.S. officials. Reports continue to circulate about Iraq's present WMD activities. After 1995 Iraq allegedly began producing chemical weapons at a facility in Sudan. In November 1997 eight Republican Guards were reportedly killed and 120 injured during an attempt to move chemical weapons out of Baghdad.[44]

Since 1991 Iraq has submitted numerous versions of its full, final, and complete declaration regarding its biological weapons program. None of these declarations has been full, final, or complete. Until July 1, 1995, Iraq claimed that its biological program was strictly defensive. On that date Iraq admitted that it had pursued industrial-scale production of aflatoxin, botulinum, and *Clostridium perfringens.* Later that month, after the defection to Jordan of Saddam's son-in-law Lieutenant General Hussein Kamal, who had run Iraq's program for procuring advanced weapons, the Iraqi regime provided significantly more detail. But UNSCOM still does not know the size of the Iraqi effort: "Expert estimates of production quantities of biological weapons agents, either by equipment capacity or by consumption of growth media, would far exceed declared amounts." UNSCOM has been unable to account for seventeen of the thirty-nine tons of growth media that Iraq imported in 1988; there is evidence that Iraq imported additional growth media in 1989 and 1990, and that material has not been accounted for either. Iraq has admitted that it removed all important equipment from Salman Pak and that none of it was destroyed by the air campaign. Iraq may be able to produce dry, powdered anthrax, which can have a long shelf life. U.S. officials believe that some 100 biological weapons facilities remain.[45]

UNSCOM is also concerned that Iraq's delivery systems, including bombs, warheads, and sprayers, still exist and are more sophisticated than previously thought. At least 25 warheads are believed to have survived the Gulf War intact. Launched against a large city under ideal meteorological conditions, each of these warheads (which are filled with anthrax and botulinum toxin) could in principle kill hundreds of thousands of people. As mentioned earlier, however, it is not clear whether Iraq has mastered the requisite technology for exploding the warheads without destroying the biological agents.[46]

Ambassador Richard Butler claims that serious gaps remain in UNSCOM's knowledge of the history, purpose, and achievements of Iraq's biological weapons program. Iraq is capable of producing the requisite equipment for manufacturing biological agents, and if UNSCOM were to stop monitoring Iraq's activities, the biological weapons program could be reactivated "almost immediately," a U.S. official reports.[47]

Cheat and Retreat, Delay and Denial

Although Iraq has officially accepted Resolution 687, its cooperation with the IAEA and UNSCOM has been sporadic. After Hussein Kamal's defection in August 1995, Baghdad admitted having organized a massive operation to deceive U.N. inspectors. During the summer of 1991 the directors of Iraq's nuclear, biological, and chemical weapons facilities were ordered to pack up "important documents" regarding the technology of production and deliver them to representatives from the Special Security Organization. The Special Republican Guard appears to have had primary responsibility for the operation, but the Intelligence Service, the Special Security Organization, the General Security Service, and Military Intelligence have also been implicated.[48]

In November 1997 Saddam expelled American weapons inspectors. U.N. inspectors attributed his action to their having come too close to parts of the biological weapons program still hidden by the Special Republican Guard, possibly at so-called presidential sites, which Iraq had forbidden UNSCOM to inspect. UNSCOM estimates (using Iraq's figures) that at that time Iraq was capable of producing some 350 liters of weapons-grade anthrax per week, sufficient to fill two warheads or four aerial bombs. Iraq keeps a close watch on the movements of UNSCOM officials, thereby establishing an early warning system for inspections.[49] The crisis nearly led to war in early 1998. Planned American military strikes were averted when U.N. Secretary-General Kofi Annan brokered a deal in which Iraq agreed to allow inspections at presidential sites provided that diplomats accompany the inspectors.

Iraq's strategy of cheat and retreat, delay and denial has led inspectors to liken implementing the U.N. resolution to peeling an onion: every time a layer is stripped away there is yet another layer. Baghdad has sacrificed more than $100 billion in oil revenues through its refusal to comply with Resolution 687 by relinquishing all WMD programs.[50]

Russia's role in UNSCOM has been problematic. UNSCOM uncovered evidence that Russia had agreed to provide Iraq with sophisticated fermentation equipment that could be used to manufacture biological weapons, although the equipment may not have been sent. It also discovered that the Russian government was spying on UNSCOM, apparently in order to warn Iraq of impending inspections so that equipment could be removed from the targeted sites.[51]

Not even the most alarmist analysts realized how advanced Iraq's WMD program was before the 1991 Gulf War. The program, a U.S. official concedes, was "very successfully hidden from the world's intelligence community." Because of inadequate intelligence, the Coalition bombing campaign against Iraq barely made a dent in Baghdad's long-term ability to produce weapons of mass destruction. Since the war Iraq has continuously tried to deceive UNSCOM, and its efforts may still be successful. "We may never know the whole story," points out Brad

Roberts, an expert in chemical and biological weapons. "The whole story may be really scary."[52]

Iraq's use of chemical weapons against its own citizens in the late 1980s demonstrated its willingness to commit acts of terrorism employing WMD. Deputy Prime Minister Tariq Aziz has denied that Iraq intends to use these weapons in terror attacks against the United States: "We are not in the business of terrorism." But Aziz warns that "there are people in other countries who are not satisfied with the situation about Iraq. If a military attack is waged against Iraq, that will increase the resentment against the Americans, and more people will be in that mood [to commit acts of terror]."[53] Despite having been defeated in the Gulf War and being monitored by UNSCOM, Iraq remains capable of supplying terrorists with weapons of mass destruction.

The case of Iraq demonstrates that no single policy will prevent WMD terrorism. Even a combination of preventive war and unprecedentedly intrusive international inspections has not destroyed Iraq's capacity to use these weapons in acts of terrorism around the world. A combination of remedies is needed, including policies designed to prevent WMD terrorism, to reduce its impact if it occurs, and to update international and domestic law to deal with emerging threats.

What Is to Be Done?

There is no single defense against this threat. Instead we must treat it as if it were a chronic disease, being constantly alert to the early symptoms and ready to employ, rapidly, a combination of treatments.

—William S. Cohen, 1997

The bad news is that most constraints against terrorists' use of weapons of mass destruction are eroding, and that the United States is particularly vulnerable. The good news is that there is much that can be done—and much that is already being done—to minimize the threat.

The U.S. government is pursuing a wide variety of strategies for preventing WMD attacks and for minimizing loss of life if attacks do occur. Those responsible for deciding how to allocate resources for combating terrorism should bear several principles in mind.

First, terrorists are more likely to use industrial poisons or chemical or biological agents than nuclear weapons, and they are more likely to disseminate radiation than to explode a nuclear device. The danger of nuclear terrorism must be addressed because it could have devastating consequences, but it should not be emphasized at the expense of the more likely threats.

Second, if terrorists or terrorist regimes do use nuclear weapons against Americans, they are more likely to employ crude delivery vehicles than sophisticated ones. Ballistic missiles are the least likely method of delivery, and yet Congress regularly allocates more money to ballistic-missile defense than the Pentagon says it can use—roughly ten times what is spent to prevent WMD terrorism. Ballistic missile defenses may turn out to be highly beneficial, but they cannot stop bombs brought into this country on cars, planes, or boats. We Americans should be asking ourselves whether this allocation of resources is efficient. Similarly, we should be wondering about the relative security we are buying by expanding NATO or building Stealth bombers, both of which were more appropriate to Cold War threats than they are to threats of terrorism, and both of which are far more expensive than the remedies proposed in this chapter. Deciding to build five fewer Stealth bombers would save an estimated $10 billion, more than enough to fund all the policies recommended here.

Third, crude chemical and biological weapons are so easy to make that strategies for prevention are unlikely to be wholly successful. But strategies for minimizing loss of life in an attack are more likely to be effective for chemical and biological weapons than for nuclear ones, since the medical consequences of some chemical and biological agents can be reversed if victims receive immediate attention. Greater resources should therefore be devoted to managing the consequences of low-technology chemical or biological attacks, including those using industrial chemicals: the hardest to prevent and the most likely to occur. Preparing to meet this threat, for example by stockpiling pharmaceuticals and training police and firefighters, could save thousands of lives.

Small changes in regulations governing law enforcement and intelligence could have a large impact. In some cases the law has not kept pace with technology. In other cases the law has not kept pace with the changing nature of the threat. The remedies proposed here would not violate citizens' First and Fourth Amendment rights and would cost taxpayers very little.

Terrorism evokes an emotional response. Terrorists aim to make a target group feel vulnerable, and they often succeed. Intense dread of weapons of mass destruction complicates the government's strategy: should the government base its response on the *perception* of risks, or on a calculation that considers all casualties equal, whatever the emotional and symbolic content of the act? In other words, should dangers that evoke disproportionate fears receive disproportionate resources?

Political vulnerability will inevitably influence the government's response. Polling shows that Americans are increasingly afraid of major acts of terrorism, those capable of causing many casualties. They are especially fearful of terrorism involving weapons of mass destruction. Politicians feel they must be seen to be doing something about the threat, even if the threat is little understood. Informed voters will be in a better position to participate in a debate about resource allocation. They also will be in a better position to influence the balance that will be struck between public safety and civil liberties.

Deterrence

The traditional response to threats involving risk and uncertainty is to persuade the adversary not to attack by maintaining a convincing deterrent. But such a strategy requires information about the adversary's motivations and capabilities. In the case of terrorism, information is sparse. The adversaries are secretive. Their motivations are changeable and their capacities unknown. The consequences of their possible actions range from minor annoyance to society-altering catastrophe. The terrorists themselves may be unable to predict the outcome of a given attack: a virus may mutate to become harmless to human beings, but it also may mutate to become more contagious or more lethal; a mistake in the manufacture of a nuclear weapon might kill the terrorists deploying it and harm no one else.

It is difficult to preempt or deter adversaries whose identities, motivations, and likely responses are unknown. It is also difficult to preempt or deter adversaries whose responses are not rational. As noted in Chapter 5, politically motivated terrorists do not appear to be "rational actors." Political terrorists have rarely achieved their putative goals, and yet politically motivated terrorism persists. This suggests that the terrorists are motivated by something other than cost-benefit calculations about how to achieve the goals of their organization. Their individual motivations and the dynamics of their groups may lead to acts of violence that are inconsistent with their purported objectives.

For ad hoc groups seeking revenge, acts of violence are likely to be expressive rather than instrumental. They do not measure success by political changes but by a horrified and hurt audience and a humiliated target government. Ad hoc groups have little to lose and are therefore hard to deter. Religious extremists, similarly, may be actively aiming for chaos or, in some cases, for martyrdom. It is hard to deter a group that is seeking to bring on Armageddon.

State sponsors, in contrast, may be possible to deter. Deterrence is vital, since terrorist states and state-sponsored terrorists are more likely than others to be proficient technologically, and therefore to be capable of killing large numbers of people. State sponsors will be most dangerous if they believe that they can camouflage their involvement, that they have nothing left to lose, or that their adversaries lack the will (or the ability) to retaliate harshly.

Deterrence requires an understanding of the adversary's preferences, including what the leader of the group or country values most. Thus deterrence requires reliable intelligence. And the existence of such intelligence should be advertised: it is important to convince state sponsors that if they support WMD terrorism their involvement will be discovered and they will be punished.

A government's deterrent posture must be carefully crafted. On the one hand, it makes sense for governments to signal their intention to respond to state-sponsored terrorist acts with massive retaliation that may even include the use of nuclear weapons. On the other hand, threatening to retaliate with nuclear weapons for acts of chemical warfare or chemical terrorism would not be proportional and might undermine efforts toward nuclear nonproliferation. An appropriate response is to threaten massive retaliation without providing details about the weapons likely to be employed.[1] And threats of retaliation should not be limited to terrorism that crosses national borders: the world cannot afford to ignore terrorist acts like Iraq's chemical attacks against its own civilians. The next attack against noncombatants might well involve biological weapons, and might be significantly more deadly than anything seen thus far.

Finally, it is important to ensure that state sponsors of terrorism never be made to feel they have nothing left to lose. As much as possible, they must be brought into the international community.

Prevention

In response to a series of deadly and highly publicized terrorist attacks at the World Trade Center, the federal building in Oklahoma City, and the Tokyo subway, updated guidelines for U.S. counterterrorism policy were issued on June 21, 1995, as Presidential Decision Directive (PDD) 39. This document addresses all forms of terrorism but singles out the use of nuclear, biological, and chemical weapons as particularly worrisome: "The United States shall give the highest priority to developing effective capabilities to detect, prevent, defeat, and manage the consequences of nuclear, biological, or chemical (NBC) materials or weapons use by terrorists. The acquisition of weapons of mass destruction by a terrorist group, through theft or manufacture, is unacceptable. There is no higher priority than preventing the acquisition of this capability or removing this capability from terrorist groups potentially opposed to the U.S."[2]

Another directive, PDD 41, addresses U.S. policy for reducing the risk of "loose nukes" from Russia—the illicit transfer of nuclear weapons or their fissile-material components to states or terrorists. Taken together, these two PDDs provide a clear mandate for preventive defense against WMD terrorism. Many vulnerabilities have yet to be addressed, however, or have received insufficient attention. Better measures are needed for preventing WMD terrorism, for minimizing loss of life if prevention fails, and for improving international and domestic laws.

Project Sapphire

Shortly after the dissolution of the Soviet Union, the government of Kazakhstan discovered a half-ton cache of highly enriched uranium (HEU) that had been left, apparently forgotten, at Ulba Metallurgical Plant. In mid-1993 the Kazakhstani government asked the United States for help in disposing of the material. When U.S. officials visited Ulba, they found the site in an alarming state of disrepair. Nearly 600 kilograms of weapons-grade uranium was minimally protected from theft. It was stored behind wooden doors secured only with padlocks. The fence surrounding the compound was rickety, and the site was poorly lit. The principal risk was theft by insiders, but Iranian agents reportedly had attempted, unsuccessfully, to purchase some beryllium—a material used in nuclear weapons and reactors—and had inquired about the low-enriched uranium also stored at the site. In addition to the large stock of beryllium and low-enriched uranium, there was enough HEU for some two dozen nuclear weapons.[3]

The White House determined that the only reliable way to secure the material was to bring it to the United States, where it could be converted to low-enriched uranium suitable for commercial power reactors. The operation would have to be carried out in complete secrecy: authorities worried that criminals or terrorists might attack the site while the technicians were preparing the material for transport. The U.S. government also worried that if antinuclear protesters heard about the shipment they might call attention to the operation, inadvertently endangering the technicians and guards at the site. An interagency group called the Tiger Team was formed to plan the highly secret mission—code-named Project Sapphire—to airlift the material from the vulnerable site in Kazakhstan to safe storage at Oak Ridge, Tennessee.

A mockup of the Ulba site was constructed at the Y-12 plant in Oak Ridge so the team could train for the mission. The team's main objective was to repackage the uranium for safe transport, and to do it fast. It took the team two months to assemble, test, and practice with the equipment they planned to take with them.[4]

In early October 1994 President Clinton signed a directive allowing the operation to proceed. Twenty-seven Department of Energy (DOE) personnel, three translators, and one liaison officer from the Department of Defense (DOD), along with 130 tons of equipment, were immediately flown to the Ulba site. Their task was complicated by the state of some of the nuclear material, which was unwieldy and hazardous to pack. Much of it was dangerously damp, and technicians had to bake it at high temperatures to remove water and residual oils. The Sapphire team worked up to fourteen hours a day, six days a week, preparing and packing the material. Some of it was pure HEU, some alloyed with beryllium, and some in the form of machining scrap. More than two metric tons of metal, powder, and scrap had to be repackaged in 1,400 stainless steel drums—each about the size of a quart of oil—which in turn were packed into 55-gallon containers.[5]

A clash of cultures created additional tensions. Kazakhstani technicians are accustomed to largely ignoring radiation hazards. For example, at the nuclear test range at Semipalatinsk not far from Ulba (which was closed in August 1991), physicists engaged in a routine that horrified their American counterparts, who are trained to put safety first. They would place a glass of vodka on the ground above the test site and after the blast would race to see who could get to the vodka first. In above-ground tests, the building that housed observers was located so close to the blast that it reportedly shook furiously.[6]

Once the operation was under way, the Tiger Team met almost daily under conditions of strict secrecy. It was crucial that they finish before the onset of winter, and that there be no leaks to the press before the material was secured. In late November the material was finally ready. Three aircraft were sent to Kazakhstan to pick up the dangerous cargo, the Sapphire team, and the air force logistics crew. The flight back to the United States took more than twenty hours and required some six midair refuelings. The officer who led the transport mission described the crew's tension: "We were sitting there in the cockpit, writing Tom Clancy novels in our heads about what would happen if we had to go down." The day before Thanksgiving the cache was finally secured at the Y-12 plant in Oak Ridge.[7] From there it was sent to the Babcock and Wilcox Naval Nuclear Fuel Division in Lynchburg, Virginia, to be converted to low-enriched uranium.

Other forgotten caches of weapons-grade material are likely to be scattered about the former Soviet Union. As discussed in Chapter 6, the government of Georgia discovered a cache of HEU and spent fuel at an obsolete nuclear reactor outside Tbilisi. The governments of Kazakhstan and Georgia acted responsibly in these two cases: they requested immediate assistance in securing weapons-grade material they knew they could not adequately protect on their own. But there are no guarantees that all stockpiles will be dealt with in this way.

In Project Sapphire, the U.S. and Kazakhstani governments responded creatively and effectively to an unexpected situation. They were able to prevent the theft or export of a cache of weapons-grade material by removing it from the site. But such opportunities are unlikely to present themselves regularly, especially if the U.S. government seems unreliable or slow to respond to requests for assistance. Not all the lessons of Project Sapphire are positive in this regard. Interagency battles—especially between the State Department, the Department of Energy, and the Department of Defense—plagued the operation from start to finish. It took several years before the U.S. government was able to agree on a source of funds for a small compensation package to Kazakhstan.

In the future it would make sense to have in place a mechanism and a source of funds to ensure that the United States is prepared to respond if other opportunities to remove vulnerable nuclear materials present themselves. If terrorists or states acquire such a cache, the cost of undoing the damage could be measured in lives as well as in dollars.

MPC&A for Nuclear Materials

The Cooperative Threat Reduction Act of 1991 (known as "Nunn-Lugar") authorized the U.S. Department of Defense to help the former Soviet states destroy WMD, store and transport weapons slated for destruction, and reduce the dangers of proliferation. Additional legislation, the Defense Against Weapons of Mass Destruction Act of 1996 ("Nunn-Lugar-Domenici"), expanded the effort, including the program for securing nuclear materials at former Soviet sites.[8]

Since 1994 the Department of Energy has been cooperating with Russia and the other newly independent states of the former Soviet Union to develop and deploy a modernized system for material protection, control, and accounting (MPC&A). Cooperation is now under way at nearly all vulnerable former Soviet sites known to the U.S. government, although agreement has not yet been reached to upgrade security at all facilities. At the current level of funding, DOE anticipates it will take until the year 2002 to complete the effort.

The program's success will depend on proper implementation of equipment and procedures. It is not enough to install portal monitors and sensors or to institute a two-man rule: procedures must be followed and equipment must be used. (The two-man rule requires that scientists and technicians enter a facility in pairs to deter theft. A DOE official told me that at one facility he visited a scientist regularly took the key home at the end of the day: the facility's managers had misinterpreted the rule entirely.) Centuries of arbitrary rule in Russia have left a disrespect for the law that is likely to take many years to expunge. DOE needs to develop an appropriate mechanism for measuring the effectiveness of its assistance program. And the Russian government needs to make MPC&A a top priority and to allocate appropriate levels of funding, as it has so far been unwilling to do.

Technical fixes can only go so far: an MPC&A system is only as good as the scientists, technicians, and guards in charge of running it. It will be years before it is possible to assess the program's success. The integrity of nuclear custodians will remain an important component of security for nuclear materials, despite the vast improvements in MPC&A systems that have already been made. If the custodians of nuclear materials are unpaid and unable to feed their families, some may find ways to circumvent the controls.

Enormous stockpiles of weapons-usable nuclear materials are one of the legacies of the Cold War. Inadequate security for these materials, as well as for plutonium-containing nuclear waste, is a worldwide problem. The problem is most acute in Russia, but many countries' systems for securing nuclear facilities are also aging and in need of repair. These problems must be addressed.[9]

Border and Export Controls

The borders of the states that once belonged to the Soviet Union are not well guarded, and may allow smuggling of materials that could be used to make WMDs. Those in the southern tier—including Armenia, Azerbaijan, Kazakhstan, Russia, and Turkmenistan—are particularly permeable.

The Baltic states are also problematic. In Latvia guards are still paid the equivalent of $20 to $30 a month—not nearly enough to live on. People become border guards in Latvia knowing that the only way they can make a living is by accepting bribes, according to a U.S. official. Officials from the Baltic states readily admit their concern that they are a transit zone between East and West for shipments of stolen goods and illegal substances. "Once you have set up the infrastructure for [exporting] contraband, you can smuggle anything, including nuclear material," a senior Latvian official told me. He fears that terrorists will eventually acquire crude nuclear weapons, in part because of weak border controls: "In our national threat estimate, the probability of our government [coming] under this threat [of nuclear terrorism] ranks higher than a foreign military attack."[10]

The Latvian official asked for advice. The solution seems pretty clear. To begin with, border guards must be paid a living wage. In addition, they must be trained to recognize illicit WMD materials. Under the Nunn-Lugar-Domenici legislation, U.S. Customs and DOD are helping the newly independent states upgrade border security by training border enforcement officers to handle dangerous materials, such as radioactive metals, and to identify weapons of mass destruction, their components, and other dual-use materials.[11]

Customs has also supplied border-enforcement agencies with vans fitted with back-scatter X-ray equipment (for detecting organic materials like narcotics and plastic explosives) and radiation-detection equipment, endoscopes (for looking into gas tanks), and more basic gear like screwdrivers and raincoats. One of these devices, the material-identification probe system, can distinguish between identical-looking metals. For example, zirconium and aluminum rods look exactly the same, but the former is a dual-use metal controlled for export. Similarly, 440 steel looks identical to maraging steel, but the latter is used in gas centrifuge rotors used for enriching uranium. In 1997 the program for upgrading border controls was expanded into all of the states of the former Soviet Union and Eastern and Central Europe.

Stopping the flow of illicit materials requires more than state-of-the-art WMD detectors. It requires honest, determined guards. And honesty among border guards appears to be highly correlated with their salaries. "The key to our ultimate success," a U.S. Customs official says, "is to demonstrate that enforcing export controls brings in revenues."

And these revenues must be used, in part, to pay guards an adequate living.

DOE is also working with Russia and the other states of the former Soviet Union countries to improve export-control legislation. Ukraine, for example, asked DOE scientists to help develop a list of materials that should be controlled. Ukraine accepted DOE recommendations, and in April 1996 it became a member of the Nuclear Supplier Group, a group of countries that agree to restrict export of dual-use technologies that could be used to make nuclear weapons. Los Alamos National Laboratory and the Kurchatov Institute in Moscow are jointly developing a computerized system for analyzing nuclear exports, to help Russia's Ministry of Atomic Energy speed up the process of reviewing export licenses.[12]

Several problems confound these programs for improving border security. One is that the former Soviet borders are so long, and the traffic is so heavy, that there is little hope of catching smugglers without the assistance of intelligence and law enforcement agencies. The FBI has a program for training law enforcement officers, but it too is plagued by difficulties. Law enforcement and border-control agencies in many of the former Soviet states are so corrupt that there is no guarantee that the training provided by U.S. agencies will not be used to enhance smuggling rather than prevent it. This is not to suggest that the effort is not worthwhile, only that results need to be continuously assessed; and that good results will be difficult and time consuming to achieve. A third problem is the need to coordinate all U.S. government programs aimed at stopping diversion or smuggling of WMD. Although the National Security Council has tried to get Customs, the CIA, DOE, and the FBI to coordinate their programs, the FBI in particular resists sharing information. These programs are too important to be conducted on an ad hoc basis; they need to be better coordinated.

Stopping the Brain Drain

Because of budget cutbacks at scientific institutes throughout the former Soviet Union, workers often go unpaid for months. Two programs, the International Science and Technology Centers and Initiatives for Proliferation Prevention (IPP), help former weapons scientists find alternative employment in civilian projects. Under the IPP program, DOE national laboratories work with their counterparts in the former Soviet Union to identify R&D projects that have commercial potential. The next step is to help the laboratories develop partnerships with U.S. businesses for these projects. The program provides seed money to reduce the risk borne by U.S. firms.

In one project scientists developed a technique to un-irradiate milk contaminated by the Chernobyl reactor, so that local children would have safe milk to drink. In another project Russian scientists are working with Harvard Medical School on a new diphtheria vaccine. Projects are now under way in materials science, biotechnology, MPC&A, and instrumentation. U.S. scientists involved in the program say it has vastly exceeded their expectations: the approach taken by their counterparts in the former Soviet Union is often "innovative and different from the U.S. approach to similar issues."[13] Nonetheless, both programs are perpetually underfunded.

Analyzing Data

U.S. government agencies maintain at least twelve databases related to WMD terrorism. But agencies rarely disseminate their findings. Often they are unaware of their counterparts' efforts, and the same data are collected twice (or more than twice). Worse still, there is no analysis of worldwide terrorism that includes both domestic and international terrorism. Efforts to share data internationally are also inadequate.

The most comprehensive analysis of trends in terrorism is performed by the CIA. But the CIA, claiming inadequate funding, has not made a serious effort to determine which attributes of terrorist groups are correlated with a propensity to commit acts of extreme violence. If the U.S. government is to have any hope of predicting and preventing WMD terrorism, these data must be analyzed properly. Moreover, the costs of analysis are minimal: it is a waste of the taxpayers' money to gather data and then leave it unanalyzed.

Minimizing Loss of Life

In a 1996 test of the government's ability to respond to nuclear terrorism, FBI Special Weapons and Tactics (SWAT) teams, Hostage Response Teams, Explosive Ordnance Disposal (EOD) troops, and members of DOE's Nuclear Emergency Search Team (NEST) fanned out in New Orleans, hunting for mock terrorists supposedly planning a nuclear attack. This was the scenario: A confidential informant had warned the FBI that members of a domestic terrorist group, Patriots for National Unity, were in the area, plotting to assemble and use homemade nuclear devices in the United States. As their opening salvo, the terrorists planned to detonate a bomb at a small airport outside New Orleans. Some of the terrorists, including the group's leader, had stationed themselves on a boat in the harbor. They had taken four hostages aboard, and were apparently prepared to defend the vessel with another nuclear device. Authorities raced against the clock to locate and disable the homemade bombs before the terrorists exploded them.[14]

In another exercise, in 1994, extortionists threatened to detonate a bomb at the Summer Olympics unless authorities gave them $3 million. To prove they had the requisite material in hand, the terrorists sent authorities a shipping label from a stolen shipment of spent nuclear fuel. When the government refused to acquiesce to their demands, the terrorists detonated a nuclear device as a demonstration and threatened to destroy the Georgia Dome (where the 1996 Summer Olympics were to be held) with a second one. In this scenario the government "won": before the terrorists were able to detonate the second device, the FBI found and arrested them.[15] But would the government prevail in real life?

The exercise revealed weaknesses. Of particular concern was the difficulty of cooperation between agencies whose priorities and incentives occasionally conflict. Critics charged that the FBI was "narrowly focused" on identifying and capturing the terrorists, while DOE and DOD were focused on disabling the bomb. Disablement was "not fully achieved by methods and procedures that would ensure adequate, accurate, or valid incapacitation of the [nuclear device] with no or minimal radioactive dispersal." NEST had "stacked the deck" in its own favor.

A report subsequently commissioned by DOE concluded that NEST's ability to deal with the full range of terrorist nuclear devices was limited and funding was inadequate.[16]

The government is similarly unprepared to deal with acts of chemical or biological terrorism. New York City ran a drill at the subway stop at First Avenue and Fourteenth Street to test preparedness to respond to a chemical incident. Monitors of the drill concluded that dozens of firefighters and police would have been killed for lack of proper communication and equipment.

Unfortunately, New York is not alone, according to the Federal Emergency Management Agency (FEMA). Most metropolitan areas are currently unprepared. The public health infrastructure is weakly coordinated; there is limited integration among local, state, and federal officials and the chain of command is not entirely clear; police and firefighters are barely trained or not trained at all to respond to nuclear, biological, or chemical incidents; and there is no reliable system for communication in case of emergency. Few emergency personnel are trained even to recognize the effects of poisoning by chemical weapons. Cities have not stockpiled pharmaceuticals, and hospitals are unprepared to deal with contaminated patients or large numbers of casualties. Hospitals are unlikely even to have enough body bags. In most hospitals around the country, victims of terrorism involving chemical, biological, or radiological agents would probably have to be hosed down in the parking lot to prevent exposing emergency room patients and personnel, according to a former director of the National Disaster Medical System. While the U.S. government is attempting to improve its ability to respond to chemical and biological incidents, its efforts to date have been "problematic" and the effectiveness of the programs is "uncertain."[17]

Nuclear Incidents

DOE's Nuclear Emergency Search Team (NEST) is composed of about 1,000 nuclear weapons scientists who have volunteered for the job. NEST is prepared to search for and disable nuclear devices hidden by terrorists in American or foreign cities. NEST is also trained to disable or contain radiation dispersal. Aircraft and equipment are positioned at airports near Las Vegas and Washington, D.C., for emergency deployment.[18]

When the government receives a threat of nuclear blackmail, NEST decides whether it is credible or a hoax. NEST has analyzed more than a hundred nuclear threats, and has deployed its teams in response to about thirty, all of which turned out to be hoaxes. Although NEST has sophisticated radiation-detection devices, their utility is limited in large urban areas. A recent assessment of NEST capabilities warns: "In reality, the probability of locating a nuclear device in a high density area is very low unless search efforts can be focused by good intelligence information." Therefore, cooperation between NEST and the FBI is essential.[19]

Highly specialized skills are required to disable nuclear weapons, and with the reduced demand for expertise in nuclear weapons, the pool of scientists with these skills is shrinking. The U.S. government knows little about how to disable Russian weapons—and yet if nuclear weapons are stolen they are most likely to be Russian. In interviews DOE officials stressed the need for more R&D on overcoming booby-traps and improving disablement technologies. Although most experts agree that NEST's ability to find hidden bombs needs to be improved, some doubt that it will be possible to increase the range of currently available detection technologies.[20]

Chemical and Biological Incidents

If a chemical or biological weapon is used without warning, as occurred in Tokyo, the front line of defense is likely to be the "first responders"—firefighters, police, and other emergency personnel. If the first responders are not trained and equipped for managing the consequences of such attacks, lives may be needlessly lost.

FEMA is used to dealing with natural disasters and having to coordinate with local officials. But an act of terrorism would create both a crime scene and a disaster—making it necessary for agencies that do not usually work together to coordinate their efforts. "Oklahoma City was a good test case," a FEMA official says, "in the sense that it revealed the competing priorities of the FBI and FEMA.

The FBI's principal objective was to preserve evidence, while FEMA wanted only to save lives."[21]

Fire departments are generally prepared to deal with hazardous materials (called HAZMAT), but chemical agents can be significantly more toxic than other hazardous materials, and biological agents are more deadly still. In a terrorist attack using chemical or biological weapons, firefighters will be just as vulnerable as the immediate victims of the attack unless they immediately don protective gear. So the crucial first step is to recognize the attack for what it is—and this requires training and detection equipment. "If a cop sees a guy crawl out of a subway station and collapse," not realizing there has been a chemical attack, his instinct will be to investigate, explains Gary Eifried, who trains first responders to deal with chemical and biological attacks. If he follows his instinct rather than remembering his training, "he's going to wind up dead."[22]

Unlike HAZMAT incidents, terrorist attacks would involve coordination not only among local first responders but also between local and federal personnel. In the case of chemical or biological attacks, the Army's Technical Escort Unit would probably be mobilized along with experts from the Army's Medical Research Institute of Chemical Defense, the U.S. Army Medical Research Institute of Infectious Diseases, and Navy medical research labs.[23]

In an attack involving biological agents, public health authorities may not realize there has been an attack until victims start arriving at hospitals, perhaps several days later, looking as though they have the flu. In many cases it is too late to treat victims once they begin showing symptoms: to be effective, pharmaceuticals must be administered before the onset of symptoms. For this reason, local authorities must be trained to respond immediately to their earliest suspicions. And local hospitals must be trained and equipped to respond to biological-agent attacks.

An official of the Office of Emergency Preparedness summarized his concerns about "narrow thinking" on the part of emergency response personnel: "We get into ruts. We worry about some ten agents, and the eleventh one is the one they will use . . . Or [the terrorists will] use a technique we haven't thought of. During the Olympics . . . everyone was concerned about airplanes flying over Atlanta and dropping something. Nobody had thought about freight trains. We allowed freight trains to pass under the site at night—somebody could easily have released something from a train."[24]

The Office of Emergency Preparedness (in the Department of Health and Human Services) is proposing to train staff members at poison control centers to field calls from emergency personnel concerned about possible use of chemical or biological agents. Questions that could not be addressed locally would be funneled to one or two poison centers where personnel were more extensively trained. Any questions that stumped the special poison centers would be funneled to the Centers for Disease Control (CDC).[25]

FEMA and DOD are organizing a training program for state and local authorities to be implemented in 120 metropolitan areas. The program is now well under way. DOD is conducting assessments of cities, participating in interagency training for first responders, testing the cities' response capability, providing an inventory of assets available to state and local governments, and creating a hotline and a website to improve the sharing of information with local and state governments.[26]

DOE laboratories are conducting case studies to enable first responders to identify their strengths and weaknesses. Models developed by the laboratories will help first responders predict the effects of chemical or biological attacks on specific targets like subways and urban areas. The national laboratories are also developing new technologies for decontamination, including formaldehyde foams, cold plasma gels, catalytic sorbents, and fluorescing agents that will identify regions of contamination—but some scientists are concerned that these technologies may not be available for decades.[27]

In April 1996, just over a year after the Aum Shinrikiyo's nerve gas attack in the Tokyo subway, the U.S. Marine Corps created a Chemical/Biological Incident Response Force, based at Camp Lejeune, North Carolina, to respond to chemical and biological incidents anywhere in the world. The Pentagon is also developing a joint rapid-response team that will combine the chemical-biological expertise of the military's branches under the Chemical Biological Defense Command at Aberdeen, Maryland. The unit, called the Chemical-Biological Quick Response Force, will consist of up to 500 personnel. These teams will be able to be pre-deployed at high-profile events likely to attract the interest of terrorists, such as the Olympics. (But chemical agents act so quickly that, in the absence of warning, local officials will have to be responsible for minimizing loss of life.)[28]

Despite these admirable efforts to improve the government's ability to minimize loss of life in the event of chemical or biological attacks, problems remain. "It is likely that the terrorist use of a weapon of mass destruction in a large metropolitan area . . . would overwhelm the capabilities of many local and state governments almost immediately," Jonathan Tucker concluded in 1997.[29] Too little attention has been paid to industrial and agricultural chemicals, anti-crop and anti-livestock agents, and crude dissemination techniques for chemical and biological agents. The government needs to increase its surveillance of diseases of public health importance in humans, animals and plants. Pharmaceuticals manufacturers should be paid to maintain an adequate stockpile of drugs to be used in case of biological attack, and the drugs should be continuously rotated to prevent them from becoming dated. For some agents, pharmaceuticals have yet to be developed; the government should facilitate relevant research. More exercises are needed to clarify the chain of command in practice (as distinct from on paper) and to ensure that emergency personnel are psychologically and physically prepared.

Another challenge in the event of a chemical or biological attack will be to prevent the public from panicking and attempting to flee. Victims of persistent chemical agents may contaminate others around them. People exposed to contagious biological agents may spread disease. The Public Health Service, in developing a strategy for dealing with nuclear, biological, and chemical weapons, must take panic into consideration—for example, by preparing public service announcements to inform the public about the nature of the threat, how to minimize exposure, and where to seek treatment or counseling.[30]

Detection Devices

In the past, biological agents were the hardest to detect of the weapons of mass destruction. Polymerase chain reaction (PCR) technology now makes possible the development of highly sensitive detectors for some biological organisms. PCR-DNA detectors identify organisms by replicating a single strand of DNA and sequencing it, a process that takes several hours. They are capable of detecting minute samples of biological agent. But the current technology is cumbersome and too time consuming to be of use in terrorist attacks or on the battlefield. The detectors also have a high false-negative rate, and will work only for certain strains. If an adversary or terrorist group were to develop a new strain or a genetically altered organism, currently available detectors might not work.

In programs funded by the Defense Advanced Research Projects Agency (DARPA) and DOE, research is under way at the national laboratories to simplify and speed up PCR sensors to make it possible to detect and identify biological agents in minutes. Challenges remain in several areas, including sample collection and preparation and the development of biological procedures to detect multiple agents simultaneously.[31]

The Navy Medical Research Institute has developed a mobile laboratory that combines immunoanalysis, PCR, and traditional microbiological procedures for rapid identification of biological agents. An immunochromatographic assay takes from five to fifteen minutes, while the PCR analysis takes twenty minutes. The entire laboratory weighs approximately 300 pounds and can be carried in four boxes.[32] While the mobile laboratory was developed for military purposes, it could be used in cases of suspected biological terrorism.

New fabrication technologies such as microelectromechanical systems and microfluidics may make possible the miniaturization of detectors, including those based on older technologies like enzyme immunoassay. Highly sensitive, tiny (palm-sized and smaller) detectors will soon be available. The U.S. Army hopes eventually to place these detectors on soldiers' dog tags.[33]

Genetic engineering makes it possible to modify pathogens to evade detection (and to complicate treatment). New detection technologies called "canaries on a chip" will detect the presence of specific classes of microorganisms, such as the genus *Bacillus,* without identifying the specific microorganism within that class. Like the real canaries used to monitor the air quality in coal mines, these devices will monitor the reactions of living cells exposed to the environment, and will provide warning to everyone in the area to seek treatment.[34]

Vaccines and Pharmaceuticals

Currently available vaccines against viral and bacterial infections are of two types: killed pathogens and live attenuated vaccines. The latter are more effective but potentially dangerous, as the vaccines themselves may mutate to become pathogenic. Moreover, live attenuated vaccines are not effective when pathogens mutate: they work only for particular strains. When a flu virus mutates, a new vaccine must be developed; similarly, if terrorists alter biological agents to produce a new strain, currently available vaccines (as well as some pharmaceuticals) will not work. There are surprisingly few vaccines available against agents likely to be used as weapons, and the vaccines are unlikely to work against genetically engineered microorganisms.

Advances in technology will make possible the development of even more dangerous agents such as aerosolized, highly infectious viruses, agents that are resistant to ultraviolet light or heat, and synthetic toxins that mimic the body's natural bioregulators. Already at least one country (Russia) is known to have developed antibiotic-resistant bacteria. But advances in biotechnology are also allowing the development of entirely new classes of antibiotic and antiviral compounds that will work against these new types of pathogens.

Scientists participating in the Human Genome Project (an international effort to map positions of the human genes and to find their chemical building blocks) are sequencing pathogens in order to develop better vaccines and pharmaceuticals. Once the pathogens are sequenced, scientists hope they will be able to inject DNA encoding a given antigen directly into the cells and tissues of the body, which will then synthesize the antigen. As DNA sequencing and computational methods become faster and less expensive, scientists will be able to identify virulent genes common to different pathogens. Knowledge of these common sequences may lead to "DNA vaccines" that will work against multiple strains. Researchers are optimistic that DNA vaccines will be significantly more effective, less dangerous, and easier to deploy than those currently available. In another exciting area of research, scientists hope to use apoptosins, the proteins that trigger programmed cell death, to trigger the death of pathogenic bacteria.[35]

In another project funded by DARPA, physical chemists are exploring the possibility of scrubbing pathogens from the bloodstream by transporting them to the liver to be destroyed. The strategy would involve using a polymer with a monoclonal antibody at each end: one antibody would target a protein on the surface of red blood cells called a complement receptor, while the other would target the specified pathogen. The group's goal is to determine whether this technique could be used as a form of passive immunization or as a therapy for acute infection.[36]

In other research, DOE is working on atmospheric-transport modeling to study the fate of chemical or biological species in the atmosphere and their transport through cities. Decontamination technologies now under development include chemical and biological decontamination foams and low-temperature plasma jets that produce a very reactive form of oxygen that decomposes chemical agents and kills biological pathogens.[37]

Strengthening Laws

How much are we willing to compromise citizens' fundamental rights—such as their privacy—in order to reduce a risk we know little about? One side of the argument is that government efforts to combat terrorism are already eroding fundamental rights and that further encroachment is unwarranted, especially if the basis for the infringement is irrational dread. The other is that terrorism has the potential to compromise citizens' rights in ways far more grievous than government infringements of their privacy. When the founding fathers framed the Bill of Rights, they probably did not foresee Aum Shinrikiyo or imagine the kinds of scientific advances that have made it possible to commit mass violence.

The best way for governments to maximize the probability of

having advance warning of terrorist attacks is to infiltrate terrorist groups. Both of these strategies entail uncomfortable trade-offs with individual rights, trade-offs with which Americans need to grapple before terrorist incidents occur.

Investigating Individuals

Strict rules govern the FBI's ability to investigate potential (as distinct from known) terrorists.

In determining whether an investigation is warranted, the FBI is instructed to consider the magnitude of the threatened harm, the likelihood that it will occur, the immediacy of the threat, and the danger to privacy and free expression posed by an investigation. Moreover, "there must be an objective, factual basis for initiating the investigation; a mere hunch is insufficient . . . In the absence of any information indicating planned violence by a group or enterprise, mere speculation that force or violence might occur during the course of an otherwise peaceable demonstration is not sufficient grounds for initiation of an investigation." Thus the FBI may not be able to investigate millenarian cults or white-supremacist organizations that are identified by experts as especially likely to engage in terrorism, or that advocate terrorist acts in their published literature, unless there is a reasonable indication that the group is carrying out or about to carry out a crime.[38]

Given domestic and international groups' growing interest in WMD,[39] the government may want to consider modifying existing rules to allow the FBI more leeway in initiating an investigation, while still protecting First and Fourth Amendment rights.

Monitoring Communications

The Department of Justice's wiretap regulations are more stringent than is necessary under the First Amendment, and the FBI has interpreted those regulations even more narrowly than the Department of Justice. A number of small changes could enhance the FBI's effectiveness without compromising civil liberties. For example, some terrorists evade the FBI by changing their phone numbers frequently. Current regulations require that the FBI seek a new warrant for every new phone number, a process that takes two days. A small change in the regulations—making the warrant apply to a person rather than a specific phone number—would help the FBI, and would be unlikely to encroach on citizens' rights.[40]

Criminals on the Internet, according to the Department of Justice, steal $10 billion a year from financial institutions in the United States, and American companies are suffering even greater losses from economic sabotage abroad. Private communications over cellular phones and over the Internet are easy to monitor and intercept. Approximately 10 billion words in computer-readable form can be intercepted for less than $1, allowing thieves or spies to find valuable information quickly and inexpensively. To protect against these dangers, companies and individuals are encrypting their messages. But encryption has a negative side: it is a boon for criminals and terrorists.[41]

Right-wing extremists and other groups are using the Internet not only to attract "more sophisticated" recruits but also to discuss grand strategy, "swap anthrax recipes," and plan operations. Some foreign extremist groups with representatives in the United States are also using the Internet. Many of these conversations are encrypted. There is a danger that encryption will become so widespread that there will be "no practical prospect of listening to the conversations of criminals or terrorists."[42]

The U.S. government supports a system called key escrow encryption, which allows a trusted third party access to encrypted data. If the government had a court order, it would be able to use recovery keys to obtain encrypted information and decode it.

The key escrow approach could facilitate monitoring of terrorists' conversations, but manufacturers of encryption software are loath to provide the government with keys. The system would only work if it were employed internationally and governments were prepared to share the information obtained. Escrowed encryption can be easily circumvented, according to the National Academy of Sciences, and this somewhat (but not completely) reduces its usefulness. Key escrow encryption is a good idea in principle, but further research is needed on how to make it work. John Deutch, a former director of the CIA, believes that a key recovery system can be designed to serve the needs of both business managers and law enforcement officials, but that "it is by no means clear that the key recovery initiative will be successful."[43]

Using Informants

Because many terrorist groups are using modes of communication that are difficult to monitor, infiltration may be the only way to gather intelligence about planned operations. CIA officials complain bitterly about new regulations that prohibit using informants who are not "squeaky clean." As a former CIA director put it, "if the point is to spy on church groups and Parent Teacher Associations," law-abiding citizens make good informants. But terrorist groups consist entirely of violators of human rights. "If we want to infiltrate Hezbollah and other terrorist groups, we're going to have to deal with killers."[44] There are compelling arguments on both sides of this debate, but in the long run these new regulations are likely to damage Americans' security.

Prohibiting Dissemination of Information

Implementing legislation required under both the Biological and Chemical Weapons Conventions makes it a crime to assist anyone—including foreign powers, commercial entities, or individuals—in the production of biological or chemical warfare agents for purposes prohibited under the conventions. One way of assisting in such activities is by providing instructions, for example through bomb-making manuals. Bomb-making and murder manuals have been used to commit crimes and are likely to continue to play "a significant role in aiding those intent upon committing future acts of terrorism and violence," according to a Justice Department study.[45]

One of the publishers of bomb-making manuals is Paladin Press, whose list includes books providing instructions for producing and using chemical agents, as well as more traditional murder manuals. In a landmark case, prosecutors argued that James Perry, a hired assassin, followed instructions in *Hit Man,* a murder manual published by Paladin, to kill his client's ex-wife, their brain-damaged son, and the child's nurse. A lawyer who felt strongly that Paladin had "aided and abetted" the murderer sued the press on behalf of the dead woman's sisters. During the trial, Paladin admitted that as part of its marketing strategy it tried to reach a broad audience, including "criminals and would-be criminals who desire information and instructions on how to commit crimes." A district court in Maryland ruled that Paladin's publication and dissemination of the book was entitled to constitutional protection, and that the First Amendment precluded liability, despite the defendants' stipulation that they "intended and had knowledge that their publication would be used, upon receipt, by criminals and would-be criminals to plan and execute the crime of murder for hire, in the manner set forth in the publication." The Department of Justice believes that the district court's First Amendment analysis was "open to question." A federal appeals court reversed the district court's ruling, holding that *Hit Man* is not protected speech but a blueprint for murder.[46]

Additional legislation could make it easier to convict murder-manual publishers who admit marketing to "criminals and would-be criminals." Existing federal law prohibits the dissemination of unclassified information related to the making of nuclear, biological, chemical, or conventional bombs, but only under narrowly specified conditions:

- *Conspiracy:* if the person disseminating the information is actively conspiring with another person to commit a federal crime and the information will be used in carrying out that crime.[47]
- *Solicitation:* if the person disseminating the information has persuaded another person, especially with financial inducements or threats, to commit a federal crime.[48]
- *Aiding and abetting:* if the person disseminating the information intends to assist in a federal crime, and if the recipient of the information actually carries out the crime. There is no law prohibiting *attempted* aiding and

abetting. Legislation of 1996 broadens the scope to include crimes that were not carried out: the disseminator may be culpable, but only if the court finds that the information provided constituted "material support or resources," and if the crime was a specified terrorist offense.[49]

- *Civil disorder:* if the person disseminating the information intends for the information to be used to create a civil disorder, and if it can be shown that publishing the information is a form of teaching or demonstration.[50]

Except in these narrow circumstances, the federal criminal code does not effectively prohibit the dissemination of unclassified bomb-making information, even when the disseminator intends that the information be used for criminal purposes or knows that the recipient plans to use it for unlawful action. Thus there is a statutory gap.

The Justice Department argues that it is not legally possible to ban bomb-making manuals outright, even when they contain (unclassified) material related to weapons of mass destruction. Moreover, if the government did ban bomb-making manuals, authors could put their ideas for terrorist operations into novels. It would be contrary to everything America stands for if the government hired book police to ensure that novels were "safe" for publication.

But narrowly drafted legislation could fill this gap without violating the First Amendment. The Justice Department has concluded that legally such legislation could apply only to cases in which publishers *intend* that the information be used to commit crimes or know that a *particular person* intends to use the information to commit crimes. Further consideration should be given to other ways to strengthen controls on these dangerous manuals.[51] The Department of Justice recommends the following language: "It shall be unlawful for any person—(a) to teach or demonstrate the making or use of an explosive, a destructive device, or a weapon of mass destruction, or to distribute by any means information pertaining to, in whole or in part, the manufacture or use of such an explosive, device, or weapon, intending that such teaching, demonstration, or information be used for, or in furtherance of, an activity that constitutes a Federal criminal offense or a State or local criminal offense affecting interstate commerce; or (b) to teach or demonstrate to any particular person the making or use of an explosive, a destructive device, or a weapon of mass destruction, or to distribute to any particular person, by any means, information pertaining to, in whole or in part, the manufacture or use of such an explosive, device, or weapon, knowing that such particular person intends to use such teaching, demonstration, or information for, or in furtherance of, an activity that constitutes a Federal criminal offense or a State or local criminal offense affecting interstate commerce."

Another approach would be to organize a public campaign—perhaps including op-ed pieces, speeches, and television appearances—to persuade publishers not to sell bomb-making manuals.

Banning Possession of Biological Agents

States that sponsor terrorism are acquiring biological weapons. But there is no mechanism for investigating suspect sites under the 1972 Biological Weapons Convention (BWC). Unlike the Nuclear Nonproliferation Treaty (NPT) and the Chemical Weapons Convention (CWC), the BWC has no protocol for inspections and no requirement to declare facilities that could be used to produce biological agents. Negotiations have been under way since 1995 to develop a compliance-and-transparency protocol. Until recently the United States has not been a strong supporter of the effort to strengthen the BWC, in part because of concerns about potential threats to confidential business and national security information, and in part out of fear that an inspection protocol could lull parties into a false sense of safety.[52]

While no inspection protocol can guarantee compliance, challenge inspections (short-notice inspections at suspect sites) can make proliferation more risky and more expensive. The U.S. government should avoid undue optimism about the effectiveness of inspections, but should nonetheless support such a protocol. Had such a protocol been in place in 1979, it would have been possible to investigate a suspicious outbreak of anthrax in Sverdlovsk, Russia, which President

Yeltsin subsequently admitted had been caused by an accidental release from a banned biological weapons facility.[53] The riskier and more expensive biological weapons programs become, the harder it will be for terrorists to acquire sophisticated delivery systems. Arms control and inspection regimes are one element of a broader set of confidence-building measures.

Other measures being considered for strengthening the BWC include adopting a protocol for investigating alleged use and adopting measures for investigating unusual outbreaks of disease. An unusual outbreak of disease might result from a release of biological agents or from the natural occurrence of an emerging or reemerging disease. Epidemiological investigations not only would help clarify questions of alleged use but also could improve public health more generally.[54] It would be useful to put in place a mechanism for monitoring unusual outbreaks of disease in animals and plants as well.

Nations that ratify the BWC are required to adopt legislation making it illegal for individuals to possess biological agents for purposes prohibited by the convention. The U.S. Congress fulfilled this requirement in the Biological Weapons Act of 1989. Congress intended the act to protect the nation from biological terrorism, but the legislation is weak. It is illegal for individuals to possess biological agents for use as weapons, but possession for "peaceful purposes," such as developing vaccines, is not forbidden.

Thus when Larry Wayne Harris bought three vials of the bacterium that causes bubonic plague, he was not breaking the law. He was convicted of misrepresenting himself in his purchase order—not of possessing an illegal substance. Moreover, as Harris explained, no law prevented him from isolating the bacteria that cause anthrax, brucellosis, or tularemia, as he claims to have done. After the 1995 Harris case the CDC tightened requirements for shipping 24 infectious agents and 12 toxins that pose significant risks to human health, including the bacteria that cause anthrax, bubonic plague, tularemia, and brucellosis. Shippers and receivers of these infectious agents now have to register with the CDC.[55] But CDC registration is required only for agents *shipped* after the new rules took effect. No law prohibits individuals from deliberately isolating biological agents from nature. And laboratories are not required to register agents they acquired before the new rules took effect.

Designating a National Coordinator

Finally, mounting an effective response to the threat of terrorism using weapons of mass destruction requires coordination between state and local officials, between state and federal agencies, and among numerous federal agencies involved in setting policy. The Defense Against Weapons of Mass Destruction Act of 1996 called for the establishment of a national coordinator for all policies related to countering WMD terrorism. The White House responded in May 1998 by designating an Office of the National Coordinator for Security, Infrastructure Protection, and Counter-Terrorism, which will be responsible for (among other things) coordinating programs for consequence management in the event of a terrorist attack employing WMD.[56] This coordinator will provide much-needed oversight of agencies that are now competing for dollars, repeating one another's efforts, and working at cross-purposes.

The government and the American people are faced with difficult questions. The only completely sure way to prevent WMD terrorism would be to implement far greater police control than the United States has ever known, at a level inconsistent with democratic institutions. The cure, in other words, might be worse than the disease. At the same time, the risk of overreaction after an attack must be borne in mind. If governments do not prepare in advance, they will be more likely to take actions the country will later regret, such as revoking civil liberties. The National Coordinator will have to weigh these competing concerns in developing policies.

Recommendations

- *Create an emergency fund.* Congress should create a special fund to be used when countries request immediate assistance in locating misplaced bombs or materials or securing vulnerable facilities, and to be used for future operations of the Project Sapphire type. In emergency situations like this, delayed action may result in needless loss of life. The fund should also support an interagency team of volunteers, whose members would be prepared to fly at a moment's notice to provide assistance at the scene of a terrorist nuclear, chemical, or biological threat, accident, or attack. The team would include members from law enforcement and intelligence in addition to technical experts, and would be deployed for incidents outside the United States. The fund should not be tied to a given year's budget request, and should amount to at least $50 million.

- *Accelerate efforts to secure Russia's warheads.* Poor security for Russian nuclear warheads is one of the most significant threats to international security at this time. Russia has requested assistance in upgrading its inventory and security systems. Through DOD's Cooperative Threat Reduction Program, the United States is assisting the Russian Ministry of Defense with a computer system to track nuclear weapons destined for dismantling and state-of-the-art fencing and sensor systems for storage sites. DOD is also working to improve screening and training for those who guard Russia's nuclear weapons. Increased funding is needed to expedite these projects.
- *Accelerate efforts to secure nuclear facilities.* Efforts to improve MPC&A at former Soviet nuclear facilities should be expanded and expedited. The Russian government, for its part, should make MPC&A a priority and should allocate adequate resources. DOE should develop a mechanism for measuring the program's success. And Congress should make similar programs available to all countries that request assistance with MPC&A for nuclear materials.
- *Protect U.S. and foreign borders.* Congress should allocate additional funds to assist former Soviet governments in improving security along their borders, especially in the southern tier. Former Soviet governments, for their part,

need to ensure that border and law enforcement officers are paid a living wage. U.S. government programs need to be better coordinated. And funding should be increased for the efforts of DOE and Customs to develop new technologies for detecting WMD at points of entry both at home and abroad.

- *Prevent brain drain.* Economic desperation in Russia's closed cities poses a grave risk to international security. Innovative programs run by DOE and the State Department help former weapons scientists find jobs in civilian projects. Rather than abandoning these efforts, as Congress regularly threatens to do, Congress and the Administration should work together with the Russian government to make these programs work better and faster. The programs should be expanded to include guards and workers in addition to scientists.[57]
- *Increase funding for R&D.* More research is needed on ways to detect, disable, and mitigate the effects of WMD. The U.S. National Laboratories have vast research capabilities that, because of inadequate funding, are not being fully exploited. To combat chemical and biological terrorism, immediate requirements include the development of better sensors and pharmaceuticals. Research on the movement of aerosols through cities and subways is also critically important. To combat nuclear terrorism, funding is needed to improve and implement nuclear forensics capabilities and to improve the detection of nuclear materials.
- *Improve international sharing of intelligence.* First, create a secure, continuously updated database of terrorist incidents, terrorists' movements, and smuggling of WMD, and make the data easily available to intelligence and law enforcement agencies around the world. Second, establish a hot line between all nuclear weapons laboratories so that nuclear weapons experts can consult one another, for example, in the event of an accidental detonation.

- *Improve domestic preparedness.* Without proper training and equipment, in the event of a WMD attack the first responders will be killed before they are able to save others. An interagency team is training the trainers of first responders, beginning in 120 cities; this training must be accelerated and expanded to other cities. Federally funded equipment and supplies should be distributed. The Metropolitan Medical Strike Teams program run by the Department of Health and Human Services should be better funded. And regular exercises are necessary to train first responders to communicate among themselves, with their counterparts in other cities, and with federal officials; and to test proposed response plans.
- *Update international and domestic laws.* Although shippers and receivers of certain deadly biological agents are required to register with the CDC, no law prohibits individuals from deliberately isolating biological agents from nature. Anyone possessing these agents should be required to register with the CDC, regardless of the agent's source. The Biological Weapons Convention should be strengthened: an inspection protocol and mechanisms for investigating unusual outbreaks of disease in humans and alleged use of biological agents should be added, and a system for monitoring unusual outbreaks of disease in plants and animals should be put in place. Congress should make it illegal to sell bomb-making manuals when publishers intend that the information be used to commit crimes or know that a particular person intends to use the information to commit crimes.

The panoply of policies discussed here can reduce the likelihood and the deadliness of WMD terrorism—if Americans recognize their importance. New technologies have made terrorism more lethal, but technology also makes terrorism easier to combat. A surprising number of remedies are easy and inexpensive. Deploying sensors and stockpiling pharmaceuticals, for example, could reduce deaths from a sophisticated biological attack from hundreds of thousands to a small fraction of that number. It makes sense to take action now, rather than after the first attack, when the response would be more emotional than measured, and when many people would be needlessly dead.

Policies to prevent WMD terrorism are unlikely to be 100 percent effective. The challenge is to make sure that if acts of WMD terrorism do occur the government is prepared to minimize loss of life, reduce public panic, and respond effectively, compassionately, and with justice.

Tables

Notes

Acknowledgments

Index

Tables

Table 1. Antipersonnel biological warfare agents

Agent	Symptoms/effects	Mortality (if untreated)	Contagiousness	Onset
Bacteria				
Bacillus anthracis (anthrax)	Cough, difficulty breathing, exhaustion, toxemia, cyanosis, terminal shock	95–100%	none (except cutaneous)	2–4 days
Yersinia pestis (pneumonic plague)[a]	High fever, headache, chills, toxemia, cyanosis, respiratory failure, circulatory collapse	100%[b]	high	2–3 days
Francisella tularensis (tularemia)	Muscle ache, chills, cough, acute respiratory distress, exhaustion, prostration	30–40%	none	2–4 days
Rickettsia				
Coxiella burnetii (Q-fever)	Fever, pains, headache	0–1%	rare	15–18 days
Virus				
Flaviviridae (Yellow fever, Dengue fever)	Headache, fever, vomiting, constipation, muscle/joint pain, bleeding, prostration, shock	rare	moderate	3–15 days
Filoviridae (Ebola virus)	Headache, fever, vomiting, diarrhea, easy bleeding, prostration, shock	50–90%	high	2–21 days

Virus *(continued)*				
Other viral hemorrhagic fevers	Headache, fever, vomiting, diarrhea, easy bleeding, prostration, shock	varying	varying	varying
Venezuelan equine encephalitis (VEE encephalomyelitis)	Joint pain, chills, headache, nausea, vomiting with diarrhea, sore throat	0–2%	low	2–5 days
Toxin				
Ricin	Weakness, fever, cough, pulmonary edema; severe respiratory distress	presumed high	none	18–24 hours
Clostridium botulinum (botulinum toxin; botulism)	Nausea, weakness, vomiting, respiratory paralysis	60–90%	none	24–36 hours
Staphylococcus aureus (staphylococcal enterotoxin B; staph infection)	Fever, headache, chills, nonproductive cough if inhaled; also nausea, vomiting, and diarrhea if ingested	0–1%	none	3–12 hours

Sources: International Institute for Strategic Studies, *Strategic Survey 1996/97* (Oxford: Oxford University Press, 1997), 32; U.S. Army Medical Research Institute of Infectious Diseases, *Medical Management of Biological Casualties* (Frederick, Md.: Fort Detrick, 1996); Frederick Sidell et al., *Jane's Chem-Bio Handbook* (Alexandria, Va.: Jane's Information Group, 1998).

a. The bubonic plague form would be caused if infected fleas were purposely released; the pneumonic form by an aerosol.

b. The 100% figure is for the pneumonic form; 50% expected for the bubonic form.

Table 2. Antiplant biological agents

Disease	Agent	Likely mode of dissemination
Viruses		
Tobacco mosaic	Tobacco mosaic virus	Aerosol
Sugar-beet curly top	Sugar-beet curly top virus	Arthropod vectors
Corn stunt	Corn stunt virus	
Hoja blanca (rice)	Hoja blanca virus	
Bacteria		
Rice blight	*Xanthamonas oryzae*	
Corn blight	*Pseudomonas alboprecipitans*	
Sugarcane wilt	*Xanthomonas vasculorum*	Aerosol
Fungi		
Late blight of potato	*Phytophythora infestans*	Aerosol/dust
Coffee rust	*Hemileia vastatrix*	Aerosol/dust
Black stem rust of cereals	*Puccinia graminis*	Aerosol/dust
Rice blast	*Pyricularia oryzae*	Aerosol/dust

Source: Charles Piller and Keith R. Yamamoto, *Gene Wars* (New York: William Morrow, 1988).

Table 3. Chemical warfare agents

Agent class	Persistence	Symptoms/effects	Rate of action
Nerve			
Tabun	low	Tightness in chest, difficulty breathing, twitching, staggering, coma, convulsions	very rapid
Sarin	low	Tightness in chest, difficulty breathing, twitching, staggering, coma, convulsions	very rapid
Soman	moderate	Tightness in chest, difficulty breathing, twitching, staggering, coma, convulsions	very rapid
VX	very high	Tightness in chest, difficulty breathing, twitching, staggering, coma, convulsions	rapid
Blister			
Sulfur mustard	very high	Burning in eyes, blistering of skin; nausea, vomiting, cardiac arrhythmia	delayed
Nitrogen mustard	high	Burning in eyes, blistering of skin; nausea, vomiting, cardiac arrhythmia	delayed
Phosgene oxime	low	Burning in eyes, blistering of skin; nausea, vomiting, cardiac arrhythmia	immediate
Lewisite	high	Burning in eyes, blistering of skin; nausea, vomiting, cardiac arrhythmia	rapid
Choking			
Phosgene	low	Irritation in eyes and throat, edema of lungs, gasping for breath	delayed
Diphosgene	low	Irritation in eyes and throat, edema of lungs, gasping for breath	variable
Chlorine	low	Irritation in eyes and throat, edema of lungs, gasping for breath	delayed
Blood			
Hydrogen cyanide	low	Confusion, dizziness, convulsions, respiration stops	rapid
Cyanogen chloride	low	Confusion, dizziness, convulsions, respiration stops	rapid
Arsine	low	Confusion, dizziness, convulsions, respiration stops	delayed

Sources: Defense Special Weapons Agency, *Weapons of Mass Destruction Terms Handbook* DSWA-AR-40H (Alexandria, Va., 1997), 54; Stockholm International Peace Research Institute, *The Problem of Chemical and Biological Warfare,* vol. 1 (New York: Humanitarian Press, 1971).

Table 4. Effects of acute whole-body exposure to radiation

Acute dose (rems)	Probable effects
0–25	No observable effects
25–100	Slight blood changes; no other observable effects
100–200	Vomiting in 5–50% within 3 hours; recovery within a few weeks, but blood-forming system may be damaged
200–600	For dose of 300+ rems, vomiting within 2 hours, loss of hair after 2 weeks; severe blood changes, hemorrhage, and infection
600–1000	Vomiting within 1 hour, severe blood changes, hemorrhage, infection, loss of hair; death in 80–100% within 2 months; for survivors, long convalescence

Source: "Fundamentals of Nuclear Energy and Nuclear Weapons Proliferation," student handbook prepared by the U.S. government.

Notes

1. Terrorism Today

1. Effects calculated with equations from Samuel Glasstone and Philip Dolan, ed., *The Effects of Nuclear Weapons* (Washington: GPO, 1977). I am grateful to Steven Fetter for assistance.

2. George Graham and Robert Peston, "Four Detained in Search for Oklahoma Bombers: US Passport-Holder Stopped at Heathrow Airport," *Financial Times,* April 21, 1995. James Vicini, "FBI Agents Escorting Blast Witness Back to US," Reuters, April 20, 1995. The Oklahoma City bomb had an approximate yield of 2.4 kilotons TNT, around 400 times less powerful than a one-kiloton nuclear weapon.

3. This would further erode the Posse Comitatus Statute, which forbids the federal government to use the army or air force to execute the law unless Congress expressly creates an exception. In 1981 the statute was amended so that the military could share intelligence and equipment training with civilian authorities, and in 1982 Congress passed legislation allowing the military to assist civilian authorities when nuclear devices are involved.

4. "NEST Consequence Report, Containment and Effects," Internal DOE memorandum, Jan. 18, 1997. The cost of cleaning up after a detonation releasing 10 kg of plutonium in a mixed urban-rural setting would be $80 billion in 1994 dollars. Douglas R. Stephens et al., "Probabilistic Cost-Benefit Analysis of Enhanced Safety Features for Strategic Nuclear Weapons at a Representative Location" (Livermore, Calif.: Livermore National Laboratory, 1994, UCRL-JC-115269).

5. U.S. Congress, Office of Technology Assessment, *Proliferation of Weapons of Mass Destruction: Assessing the Risks* (Washington: GPO, 1993, OTA-ISC-559), 54, 53.

6. In *Brandenburg v. Ohio,* 395 U.S. 444 (1969), the Supreme Court held that it is permissible under the First Amendment to advocate the use of force in violation of the law, except in cases where the advocacy is "directed to inciting or producing imminent lawless actions and is likely to incite or produce that action."

Cited in U.S. Department of Justice, "Report on the Availability of Bomb-making Information, the Extent to Which Its Dissemination Is Controlled by Federal Law, and the Extent to Which Such Dissemination May Be subject to Regulation Consistent with the First Amendment to the United States Constitution," Submitted to the U.S. House of Representatives and the U.S. Senate, April 1997, 29.

7. Samuel Huntington, "The Clash of Civilizations?" *Foreign Affairs,* Summer 1993, 25. William S. Cohen, "In the Age of Terror Weapons," *Washington Post,* Nov. 26, 1997, A19.

8. Huntington, "Clash of Civilizations," 25.

9. My regression is based on data on both domestic and international terrorist incidents worldwide. The data are from Pinkerton Risks International; I thank Eugene Mastrangelo and Frank Johns for providing them. For regression on incidents, R^2 is .89; for regression on fatalities, R^2 is .77. Organizations use various definitions of terrorism and thus report different numbers of incidents per year.

10. Raphael F. Perl, "Terrorism, the Future, and US Foreign Policy," *CRS Issue Brief,* Feb. 16, 1996.

11. U.S. Department of State, *Patterns of Global Terrorism 1995* (Washington, April 1996), 24, 27. John Lancaster, "Iran Has Strong Links to Anti-West Terror," *Washington Post,* Nov. 1, 1996, A28. Perl, "Terrorism, the Future." "Alleged Bankroller of Islamist Extremists Surfaces in Afghanistan," *Compass Newswire,* July 11, 1996. Benjamin Weiser, "Trial Begins for Chief Suspect in Trade Center Blast," *New York Times,* Aug. 4, 1997, 16. Roberto Suro, "2 Terrorist Groups Set up US Cells, Senate Panel Is Told," *Washington Post,* May 14, 1997, 4.

12. *Patterns of Global Terrorism 1995,* iii. Pierre Thomas, "A New Strain of Terrorism: Groups Are Fast, Loose, Hard to Find," *Washington Post,* Aug. 3, 1993. The groups apprehended so far have drawn on militant radicals trained in Afghani camps or veterans of the Afghan war. In 1995 authorities uncovered a conspiracy to bomb 12 American airliners within 48 hours, led by Ramzi Youssef, the convicted mastermind of the World Trade Center bombing. Ed Vulliamy and Ian Black, "The TWA 800 Disaster: FBI Follows Twists in Islamist Terror Trail," *Observer,* July 23, 1996, 18. Weiser, "Trial Begins for Chief Suspect."

13. Bruce Hoffman, "Viewpoint: Terrorism and WMD: Some Preliminary Hypotheses," *Nonproliferation Review,* Spring–Summer 1997, 48. The fraction classified as religious would be significantly larger if political groups that are partly motivated by religious conviction (such as the IRA or the Tamil Tigers) were counted. Hoffman provides slightly different numbers in *"Holy Terror": The Implications of Terrorism Motivated by a Religious Imperative* (Santa Monica: Rand, 1993, P-7834).

14. Peter Benesh, "Terrors of the Earth," *Pittsburgh Post Gazette,* Sept. 28, 1995, quoting George Joffe of the London School of Geopolitics and Josef Horchem of the International Group for Research and Information on Security.

15. E.g., Klanwatch, "Two Years After: The Patriot Movement since Oklahoma City," *Intelligence Report* (Spring 1997). Prepared statement of John O'Neill, U.S. Congress, Senate, Committee on Governmental Affairs, Permanent Subcommittee on Investigations, *Hearings on Global Proliferation of Weapons of Mass Destruction,* 104th Cong., 1st sess., pt. 1, Nov. 1, 1995, 241. Leonard Cole, "The Specter of Biological Weapons," *Scientific American* 275, no. 6, Dec. 1996 (Larry Wayne Harris only received probation). "Arkansas Man Charged for Transporting Poison," *Washington Post,* Dec. 23, 1995, A8. Jim Brooks, *Arkansas Democrat-Gazette,* Dec. 22, 1995. A particle of ricin the size of a grain of salt is enough to kill a person, and a gram is enough to kill 1,600 people.

16. Hoffman, "Viewpoint," 48. Jim Wolf, "U.S. Faces Unprecedented Terror Threat: CIA Nominee," *Reuters Newswire,* May 6, 1997. Tenet, prepared statement, Feb. 5, 1997. St. Andrews is maintaining the database started by Rand.

2. Definitions

1. Virginia Held argues that not all terrorists aim to terrify: other objectives include gaining concessions, obtaining publicity, or provoking repression. But terrorists attempt to achieve these objectives with the instrument of dread. Held, "Terrorism, Rights, and Political Goals," in R. G. Frey and Christopher W. Morris, eds., *Violence, Terrorism, and Justice* (New York: Cambridge University Press, 1991).

2. It is the means of terrorism that are morally unique, not its ends. "It would be naive to pretend that terrorism is an innocuous term, but ends must be separated from means in politics." Martha Crenshaw, *Terrorism and International Cooperation* (New York: Institute for East West Security Studies, 1989), 5. This is what I attempt to do: to define terrorism as a technique that can be used in the service of good or evil.

3. The U.S. Catholic Bishops, referring to nuclear weapons, have called indiscriminate weapons intrinsically immoral. U.S. Catholic Bishops, "The Challenge of Peace," *Origins* 13, no. 1 (May 19, 1983). This also applies to biological weapons and (to a somewhat lesser extent) to chemical weapons. But it is possible to use these weapons in a discriminative fashion. Terrorists might inject poison into individual victims, using the agent as a tool of assassination rather than a weapon of mass destruction. A sophisticated nuclear power might deliver miniature nuclear weapons far from population centers. But a crude nuclear device lacking precise guidance systems is an indiscriminate area weapon. Steven Lee, "Is the Just War Tradition Relevant in the Nuclear Age?" *Research in Philosophy and Technology* 9, no. 85.

4. But keep in mind that terrorists are far more likely to use such weapons in small-scale incidents, and that the next major regional conflict might involve significantly more casualties than did the Gulf War.

5. Yonah Alexander, "Foreword," in Donald Hanle, *Terrorism: The Newest Face of Warfare* (McLean, Va.: Pergamon Brassey's, 1989), ix. Title 22 of the United States Code, Section 2656f (d). U.S. Department of State, *Patterns of Global Terrorism 1991* (Washington, 1991).

6. Held notes that it is often impossible categorically to differentiate noncombatants from legitimate targets: in many countries children are forced to perform military service. Held, "Terrorism, Rights." Moreover, civilians are often producing armaments: "Whereas wars once affected merely the fighting men who make up a small percent of the total population, today almost the entire population of a belligerent nation becomes engaged in wartime industries." Russel H. Ewing, "The Legality of Chemical Warfare," *American Law Review* 61 (Jan./Feb. 1927), 58.

7. General Marshall allowed John P. Sutherland to record a conversation about World War II provided that the general's words not be published while he lived. Sutherland, "The Story General Marshall Told Me," *U.S. News and World Report* 4, no. 18 (Nov. 2, 1959), 53, 52. On Marshall's initial misgivings about targeting noncombatants see Barton Bernstein, "Eclipsed by Hiroshima and Nagasaki: Early Thinking about Tactical Nuclear Weapons," *International Security* 15, no. 4 (Spring 1991).

8. Descriptions of the three groups based on David C. Rapoport, "Fear and Trembling: Terrorism in Three Religious Traditions," *American Political Science Review* 78, no. 3 (Sept. 1984).

9. Ibid., 669. This is not to imply that the Zealots were responsible for 2,000 years of anti-Semitism, only that their activities had long-term effects.

10. Martin Miller, "The Intellectual Origins of Modern Terrorism," in Martha Crenshaw, ed., *Terrorism in Context* (University Park: Pennsylvania State University Press, 1995), 28.

11. Anna Geifman, *Thou Shalt Kill* (Princeton: Princeton University Press, 1993), ch. 1. Philip Pomper, "Russian Revolutionary Terrorism," in Crenshaw, ed., *Terrorism in Context,* 78. Richard Pipes, *Russia under the Old Regime* (1974; New York: Penguin, 1995), 297.

12. Miller, "Intellectual Origins of Modern Terrorism."

13. Paul Sigmund, ed. and trans., *St. Thomas Aquinas on Politics and Ethics* (New York: Norton, 1988), 65.

14. Charles Francis Adams, Address to the 13th Annual Dinner of the Confederate Veterans Camp of New York, Jan. 26, 1903, quoted in John Ellis van Courtland Moon, "Laws of War in Relation to Poisoned Weapons" (manuscript). William Tecumseh Sherman, *Memoirs* (New York, 1875), 119–120.

15. Michael Walzer, *Just and Unjust Wars* (New York: Basic Books, 1977), 36. James Turner Johnson, *Can Modern War Be Just?* (New Haven: Yale University Press, 1984), 25. One difficulty with the *jus ad bellum* criteria is that they are subjective. Who is the "right authority" if the leader is a tyrant? Who determines

who has started the war? Many wars end with both sides accusing the other of having been the aggressor. Preventive war would seem to be just, but who is to determine what behavior warrants preventive war, or who has provoked whom?

16. The earliest attempts to define a law of war proscribed targeting civilians. In 1625 Hugo Grotius referred to the need to protect civilians. The Lieber Code of 1863 declares that "the unarmed citizen is to be spared in person, property, and honor as much as the exigencies of war will admit." "Indiscriminate" attacks, which "are not directed at a specific military objective" or employ "a method or means of combat which cannot be directed at a specific military objective," are prohibited by the Protocol Additional to the Geneva Conventions of Aug. 12, 1949, as well as by subsequent treaties such as the Fundamental Rules of International Humanitarian Law Applicable in Armed Conflicts. Adam Roberts and Richard Guelff, eds., *Documents on the Laws of War* (Oxford: Clarendon, 1989).

17. Lee, "Is the Just War Tradition Relevant," 85. But modern military conflict inevitably puts noncombatants at risk. Under the double effect rule, as Michael Walzer explains, the two outcomes (the good of winning the battle or the war; and the evil of killing noncombatants) are permissible only when the actor's intentions are good and the following three conditions are met: the actor must aim narrowly at the legitimate military target; killing noncombatants must neither be his ends nor a means to reach his ends; and he must be cognizant of and seek to minimize the evil of the act.

18. Frey and Morris, *Violence, Terrorism, and Justice*, 6.

19. Ibid. An objection to utilitarianism (a version of consequentialism) in its purest form is that it would sanction murder if the only way to prevent ten murders is to commit one. Samuel Scheffler, "Introduction," in Scheffler, ed., *Consequentialism and Its Critics* (Oxford: Oxford University Press, 1988).

20. Jan Narveson, "Terrorism and Morality," in Frey and Morris, eds., *Violence, Terrorism, and Justice.*

21. A disgruntled employee's attack could be considered terrorism for the purpose of exacting revenge. The Indian government has not taken a firm stance on the cause of the chemical leak perhaps because safety features at the plant were substandard from Union Carbide's point of view. But Union Carbide's conclusions were confirmed by an independent assessment conducted by the consulting firm Arthur D. Little and in a documentary film. Interview with former Union Carbide lawyer, July 6, 1998; Jackson B. Browning (Retired Vice-President, Health, Safety and Environmental Programs at Union Carbide), "Union Carbide: Disaster at Bhopal," http://www.bhopal.com/Jbrowning.html.

22. U.N. General Assembly, *Report of the Secretary-General on Chemical and Bacteriological (Biological) Weapons and the Effects of Their Possible Use*, U.N.G.A. A/7575, 1969, 6.

23. Ibid.

24. Ibid.

25. Elinor Langer, "Chemical and Biological Warfare (I): The Research Program," *Science* 155 (Jan. 13, 1967), 176.

26. Bioregulators could also be used as weapons. These are small proteins that regulate bodily functions like sleeping or blood pressure.

27. Other anti-plant agents considered by the U.S. military include sclerotium rot (caused by a fungus), eelworm disease (caused by a microscopic worm), and curly top disease (caused by a virus)."Report on Diseases and Insect Pests of Plants as Military Weapons," declassified secret memorandum to Brigadier General W. B. Smith, Secretary, Joint U.S. Chiefs of Staff, Aug. 19, 1942. National Archives, RG 218.

28. Julian Robinson, *The Rise of CB Weapons,* vol. 1 in the series *CB Weapons Today: The Problem of Chemical and Biological Warfare* (New York: Humanities Press, 1973), 65–68.

29. Patrick Tyler, "Germ War, a Current World Threat, Is a Remembered Nightmare in China," *New York Times,* Feb. 4, 1997, 6. Tricothecene mycotoxins are alleged to have been used against the Hmong refugees fleeing Kampuchea in the late 1970s and 1980s, but a number of scholars have claimed that the "yellow rain" that fell on the refugees was actually bee feces. This case has not been resolved. Julian Robinson, Jeanne Guillemin, and Matthew Meselson, "Yellow Rain: The Story Collapses," *Foreign Policy,* no. 68 (Fall 1987).

30. *Report of the Secretary-General.*

31. U.S. Department of Energy, *Nuclear Terms Handbook 1996* (Washington, 1996).

32. Bill Sutcliffe et al., in "A Perspective on the Dangers of Plutonium" (Lawrence Livermore National Laboratory, UCRL-JC-118825), write that to be inhaled the particles must be less than 3 microns in diameter, but the toxicologist Eileen Choffnes claims that the most dangerous size is between 1 and 5 microns. An estimated .05% of the oxidized plutonium from a fire would be respirable, compared with some 20% of explosively dispersed plutonium. Ibid.

33. "Military Use of Radio-Active Materials and Organization for Defense," Bush-Conant Files, RG 227, folder 157, National Archives.

34. This section is a summary of "Fundamentals of Nuclear Energy and Nuclear Weapons Proliferation," a student handbook prepared by the U.S. government.

35. U.S. and Russian bomb designers typically use weapons-grade fissile material to ensure high reliability and efficiency of their bombs, and to allow for bombs of a size easily transported in planes or missile warheads. Uranium enriched to between 20 and 90% U-235 can also be used in a weapon, although the weapon will be proportionately heavier and larger.

36. The weapon dropped on Hiroshima, called Little Boy, was a gun-assembly device with a 12.5-kiloton yield. The weapon dropped on Nagasaki, called Fat Man, was an implosion device with a 22-kiloton yield.

3. Trojan Horses of the Body

1. Louis Renault, "War and the Law of Nations," *American Journal of International Law* 9 (1915), 1, referring to "poison or poisoned weapons."

2. Between 70,000 and 80,000 people were killed and over 70,000 injured by the bomb dropped on Hiroshima; between 35,000 and 40,000 killed and 40,000 injured in Nagasaki. Two days of conventional fire-bombing raids in Tokyo left 225,000 dead, according to Lieutenant Colonel Dave Grossman, *On Killing* (Boston: Little, Brown, 1995), 101. British bombing of German cities killed some 300,000 people, mostly civilians, and seriously injured some 780,000. (While there is no consensus among military historians regarding the exact numbers killed during conventional fire-bombing raids, there is no doubt that significantly more people died from these raids than from the two nuclear attacks.) Barton Bernstein says policymakers had no moral scruples about using the bomb; instead, they were excited that the explosion would be visible from far away, and "looked forward to the A-bomb's international political benefits—intimidating the Soviets." Bernstein, "Understanding the Atomic Bomb and the Japanese Surrender," *Diplomatic History* 19, no. 2 (Spring 1995), 236.

3. Groves to Chief of Staff, July 30, 1945, TS Docs no. 5, MED Records. Groves claimed some twenty years later: "The optimum height of burst was entirely governed by the explosive force. If the altitude of burst we used was below or too high above this optimum, the area of effective damage would be reduced." But he also said radiation was a consideration: "I had always insisted that casualties resulting from direct radiation and fallout be held to a minimum. After the Alamogordo test, when it became apparent that the burst could be many hundreds of feet above the ground, I became less concerned about radioactive fallout from too low a burst." Leslie R. Groves, *Now It Can Be Told: The Story of the Manhattan Project* (New York: Harper, 1962), 269. Bernstein says Oppenheimer and others favored a high-altitude burst, not to minimize radioactive fallout, but to maximize the blast's effects on structures. On Groves's "sustained effort to minimize evidence, and possibly to engage in self-denial, on the question of whether radiation had killed many Japanese," see Barton Bernstein, "Doing Nuclear History" *Society for Historians of American Foreign Relations Newsletter,* Sept. 1996, 33, 35.

4. Michael Mandelbaum, *The Nuclear Revolution* (New York: Cambridge University Press, 1981), 35.

5. Leon Friedman, ed., *The Law of War: A Documentary History* (New York: Random House, 1972), 38–40, citing Grotius.

6. Several scholars have contributed greatly to our understanding of the development of norms against WMD. See Jeffrey Legro, "Which Norms Matter? Revisiting the 'Failure' of Internationalism," *International Organization* 51, no. 1 (Winter 1997); Richard Price and Nina Tannenwald, "Norms and Deterrence: The Nuclear and Chemical Weapons Taboos," in Peter Katzenstein,

ed., *The Culture of National Security* (New York: Columbia University Press, 1996).

7. My account of your risky day is based on Larry Laudan, *The Book of Risks* (New York: Wiley, 1994).

8. Amos Tversky and Daniel Kahneman, "Rational Choice and the Framing of Decisions," in David E. Bell, Howard Raiffa, and Amos Tversky, eds., *Decision Making* (New York: Cambridge University Press, 1988). Robert J. Shiller, "Human Behavior and the Efficiency of the Financial System," Cowles Foundation for Research in Economics at Yale University, Feb. 1998.

9. Stephen Breyer, *Breaking the Vicious Circle* (Cambridge, Mass.: Harvard University Press, 1993), 35.

10. Mandelbaum, *Nuclear Revolution,* 39, 48. Price and Tannenwald, "Norms and Deterrence," 123.

11. See Mary Douglas, *Purity and Danger: An Analysis of the Concepts of Pollution and Taboo* (1966; New York: Routledge, 1996).

12. I am using *fear* to mean a judgment-based, general concern—a perception of risk—not an emotion-based fear of a specific thing. See Pamela Wilcox Rountree and Kenneth Land, "Perceived Risk versus Fear of Crime," *Social Forces* 74, no. 4 (June 1996).

13. Paul Slovic, Baruch Fischoff, and Sarah Lichtenstein, "Facts and Fears: Understanding Perceived Risk," in Richard Schwing and Walter Albers, eds., *Societal Risk Assessment: How Safe Is Safe Enough?* (New York: Plenum, 1980), 183. There were 11 terrorist killings in Belfast in the first six months of 1993, and 230 murders in Washington. "Degrees of Terror," *Economist,* July 10, 1993, 24.

14. A. Tversky and D. Kahneman, "Judgment under Uncertainty: Heuristics and Biases," *Science* 185 (1974). Slovic, Fischoff, and Lichtenstein, "Facts and Fears." People also tend to be overconfident about the accuracy of their assessments, even when those assessments are based on nothing more than guessing. And people seem to desire certainty: they respond to uncertainty by ignoring uncertain risks, and by believing that while others may be vulnerable they themselves are not.

15. Slovic, Fischoff, and Lichtenstein, "Facts and Fears." Paul Slovic, "Perception of Risk," *Science* 236 (1987), 280–281.

16. Carleton Savage to Paul Nitze, Oct. 24, 1951, 12, National Archives, Record Group 59, Department of State, Records of the Policy Planning Staff, 1947–1953, Box 5.

17. Slovic, Fischoff, and Lichtenstein, "Facts and Fears," 191.

18. William Miller, *The Anatomy of Disgust* (Cambridge, Mass.: Harvard University Press, 1997), 26.

19. Alberico Gentili, *De Iure Belli Libri Tres* (1612), trans. John Rolfe (Oxford: Clarendon, 1933), 155. *Hamlet* I.5.

20. Miller mentions only incest in this category; I think a taboo against murder is also maintained by disgust.

21. John of Salisbury cited in Gentili, *De Iure Belli Libri Tres,* 156. Monsieur de Vattel, *The Law of Nations* (1758), trans. Joseph Chitty (Philadelphia, 1863), 360.

22. Otto Pollak, *The Criminality of Women* (Philadelphia: University of Pennsylvania Press, 1950), 16. Margaret Hallissy, *Venomous Woman: Fear of the Female in Literature* (New York: Greenwood, 1987), xi. Dominik Wujastyk, *The Roots of Ayurveda* (New Delhi: Penguin, 1998), 123.

23. Hallissy, *Venomous Woman,* xiv.

24. Gentili, *De Iure Belli Libri Tres,* 155 and ch. VI.

25. Ibid., 157. Hallissy, *Venomous Woman,* 5.

26. Pollak, *Criminality of Women,* 11. Interview with Larry Wayne Harris, Jan. 23, 1998.

27. Teresa Brennan, *Jenseits der Hybris: Bausteine einer Neuen Okonomie* (Frankfurt am Main: Fischer, 1997). Interview with Brennan, Nov. 16, 1997.

28. Frederick R. Sidell, "Nerve Agents," in *Medical Aspects of Chemical and Biological Warfare* (1997), pt. 1 of General Russ Zajtchuk, ed., *Textbook of Military Medicine* (Washington: Office of the Surgeon General, 1989–1997). A crude kind of nerve agent, the Calabar bean, was used by native tribesmen in Western Africa as an "ordeal poison" in witchcraft trials. Calabar beans contain physostigmine, a mild cholinesterase-inhibiting (nerve) agent. W. Davis, *The Serpent and the Rainbow* (New York: Simon and Schuster, 1985), 36–37; I. H. Phillippens, B. Olivier, and B. P. Melchens, "Effects of Physostigmine on the Startle in Guinea Pigs: Two Mechanisms Involved," *Pharmacological Biochemical Behavior* 58, no. 4 (Dec. 1997).

29. When troops were insufficiently trained in gas defense, or poorly disciplined, "toxic chemicals opened the way for what might have been a decisive operation by producing casualties but more significantly, by inducing panic." Dorothy Kneeland Clark, *Effectiveness of Chemical Weapons in World War I,* Staff Paper ORO-SP-88 (Bethesda: Johns Hopkins University, 1959), DTIC AD-233081, 134. Some analysts argued at the time that CW were ideal: "Various forms of gas . . . make life miserable or vision impossible to those without a mask. Yet they do not kill. Thus instead of gas warfare being the most horrible, it is the most humane." Amos A. Fries and Clarence J. West, *Chemical Warfare* (New York: McGraw Hill, 1921.)

30. Augustin M. Prentiss, *Chemicals in War* (New York: McGraw Hill, 1937), 649. The U.S. Army Chemical Warfare Service reports a mortality rate for casualties from conventional armaments of only 8%. United States Chemical Warfare Service, Office of the Chief, *The Story of Chemical Warfare* (Washington: GPO, 1939). To put these rates into context, I have calculated the percentage of casualties that resulted in death for seven wars before World War I: Germany at war with Denmark, 1864 (34%); Germany at war with Austria, 1866 (38%); Germany at war with France, 1870 (32%); U.S. at war with Spain, 1898 (18%); England in Boer war, 1899 (26%); Japan at war with Russia, 1904 (28%); Russia at war with Japan,

1904 (18%). Based on data from A. A. Roberts, *The Poison War* (London: Heinemann, 1915).

31. Full data are not available on the Iran-Iraq War. Medical workers estimated a mortality rate under 5%, despite the use of nerve agents against relatively unprotected troops. Frederick R. Sidell and David R. Franz, "Overview: Defense against the Effects of Chemical and Biological Warfare Agents," in *Medical Aspects of Chemical and Biological Warfare.* Another source reports that of the 27,751 Iranian casualties of chemical warfare before April 8, 1987, when the Iraqis began to attack cities, only 262 died. Seth Carus, "Chemical Weapons in the Middle East," *Policy Focus: The Washington Institute for Near East Policy,* no. 9, Dec. 1988. Investigators from Physicians for Human Rights concluded that Iraq used mustard and at least one other chemical agent against the Kurds. "Winds of Death: Iraq's Use of Poison Gas against Its Kurdish Population," report by the Physicians for Human Rights. Interview with UNSCOM official Charles Duelfer, Jan. 1, 1997.

32. Nancy Kraus, Torbjorn Malmfors, and Paul Slovic, "Intuitive Toxicology: Expert and Lay Judgments of Chemical Risks," *Risk Analysis* 12, no. 2 (1992), 215.

33. Richard Preston, *The Hot Zone* (New York: Random House, 1994). Leonard Cole, *The Eleventh Plague* (New York, Freeman, 1997). Professor R. Swanepoel, communication to ProMED-mail, Jan. 31, 1997. Interviews with doctors and researchers at the U.S. Army Medical Research and Materiel Command, Jan. 17, 1997.

34. The government biologist asked not to be identified. Ed Regis, "Pathogens of Glory," *New York Times Book Review,* May 18, 1997, 18.

35. Procopius, *De Bello Persico,* quoted in Hans Zinsser, *Rats, Lice and History* (1933; Boston: Little, Brown, 1963), 146. Hallissy, *Venomous Woman,* 23.

36. Zinsser, *Rats, Lice and History,* 119–121, 125–127, 154.

37. Ibid., 80. Vincent J. Derbes, "DeMussia and the Great Plague of 1348: A Forgotten Episode of Bacteriological Warfare," *JAMA* 196, no. 1 (April 4, 1966). The biological attack was only partly responsible for the spread of plague; see George W. Christopher, Theodore J. Cieslak, Julie A. Pavlin, and Edward M. Eitzen Jr., "Biological Warfare: A Historical Perspective," *JAMA* 278, no. 5 (Aug. 6, 1997).

38. John W. Powell, "A Hidden Chapter in History," *Bulletin of the Atomic Scientists* 37, no. 8 (Oct. 1981).

39. Matt Meselson calculated that less than a gram of anthrax spores killed the 96 people, as well as sheep grazing up to 30 miles away. Nicholas Wade, "Germ Weapons: Deadly but Hard to Use," *New York Times,* Nov. 21, 1997, 17. Others believe that several kilograms of anthrax spores were released.

40. Paul Watson, "Taiwan Facing Economic Crisis as Pigs Wiped Out," *Toronto Star,* April 2, 1997, A13. Nearly 350,000 had died of AIDS in the United States as of June 1996: 189,004 blacks and 256,461 whites (non-Hispanic). Center for Disease Control, Document 320200.

41. Declassified top secret memorandum for General Groves, Summary of Target Committee Meetings on 10 and 11 May 1945, describing a meeting with Dr. Oppenheimer on May 10, 1945, National Archives, M1109, File 5 (D) (G), roll 1. Joint Chiefs of Staff, "Statement of Effect of Atomic Weapons on National Security and Military Organization," JCS 1477/5, Jan. 12, 1946 (declassified top secret), National Archives, RG 218, Box 222.

42. On the effects of fire see Lynn Eden, *Constructing Destruction: Organizations, Knowledge, and Nuclear Weapons Effects* (Ithaca: Cornell University Press, forthcoming).

43. The presence of people suffering from "intractable burns and other sustained, visible injuries" was also judged to be highly debilitating emotionally. Savage to Nitze, Oct. 24, 1951, 5.

44. Ibid., 12. *In Time of Emergency: A Citizen's Handbook on Nuclear Attacks and Natural Disasters* (Washington: Office of Civil Defense, 1968), 13–14.

45. Harry Waters, "TV's Nuclear Nightmare," *Newsweek* 102 (Nov. 21, 1983). D. M. Mayton, "Measurement of Nuclear War Attitudes: Methods and Concerns," *Basic and Applied Social Psychology* 7 (1988). R. T. Schatz and S. T. Fiske, "International Reactions to the Threat of Nuclear War: The Rise and Fall of Concern in the Eighties," *Political Psychology* 13, no. 1 (1992).

46. Slovic, Fischoff, and Lichtenstein, "Facts and Fears"; Slovic, "Perception of Risk."

47. Joseph Nye, Philip Zelikow, and David King, eds., *Why People Don't Trust Government* (Cambridge, Mass.: Harvard University Press, 1997).

48. Ernest May, "The Evolving Scope of Government," ibid.

49. Carl Builder, "Is It a Transition or a Revolution?" *Futures* 25, no. 2 (March 1993). Builder, "Peering into the Future," talk presented to the Industrial College of the Armed Forces, Feb. 8, 1996.

50. U.S. Congress, Office of Technology Assessment, "The Regulatory Environment for Science: A Technical Memorandum" (Feb. 1986), 130–132, cited in Charles Piller, *The Fail Safe Society* (New York: Basic Books, 1991), 5.

51. Piller, *Fail Safe Society.*

52. E.g., the government provided Eastman Kodak Company with maps and forecasts of potential radioactive contamination from atomic tests in the 1950s, but did not inform milk producers or the public. Radioactive iodine (I-131) concentrates in grazing cattle and is absorbed readily through milk. Thyroid cancer rates are expected to rise by 20% in those exposed. Curt Suplee, "40 Years Later, Bomb Test Fallout Raises Health Alarm," *Washington Post,* Oct. 2, 1997, 3. Hanford, a facility in Richland, Washington, which produced plutonium for DOE until the late 1980s, released several hundred curies of I-131 into the atmosphere to study its dispersion and transport characteristics. Barton J. Bernstein, "An Analysis of Two Cultures," *Public Historian* 12, no. 2 (Spring 1990), 93.

53. Kristin Shrader-Frechette, "Science versus Educated Guessing: Risk Assess-

ment, Nuclear Waste, and Public Policy," *BioScience* 46, no. 7 (July–Aug. 1996), 498.

54. Cole, *Eleventh Plague,* 214. John Moon, "Controlling Chemical and Biological Weapons through World War II," in Richard Dean Burns, ed., *Encyclopedia of Arms Control and Disarmament,* vol. 2 (New York: Scribner, 1993), 673.

4. Getting and Using the Weapons

1. Prepared statement of Edward Eitzen, U.S. Congress, Senate, Committee on Governmental Affairs, Permanent Subcommittee on Investigations, *Hearings on Global Proliferation of Weapons of Mass Destruction,* 104th Cong., 1st sess., pt. 1, Oct. 31, 1995, 112. More likely, doctors would observe large numbers of cases that resembled the flu. By the time doctors realized their patients had been victims of an attack, it would be too late to save their lives.

2. Cohen, speech to the Conference on Terrorism, Weapons of Mass Destruction, and U.S. Strategy, University of Georgia, April 28, 1997. Prepared statement of George Tenet, U.S. Congress, Senate, Select Committee on Intelligence, *Hearing on Current and Projected National Security Threats to the United States,* 105th Cong., 1st sess., Feb. 5, 1997.

3. "Albania: Army Officer Urges Return of Looted Chemical Weapons," FBIS-EEU-97–096, transcribed text, April 6, 1997. Source: Paris AFP in English 1740 GMT April 6, 1997.

4. Agence France Presse, "Russian Security Inadequate for Chemical Weapons Storage," Aug. 2, 1995.

5. Prepared statement of Vil Mirzayanov, U.S. Senate, *Hearings on Global Proliferation of WMD,* Nov. 1, 1995. A. Cooperman and K. Belianinov, "Moonlighting by Modem in Russia," *U.S. News and World Report,* April 17, 1995. James Adams, "Gadaffi Lures South Africa's Top Germ Warfare Scientists," *Sunday Times,* Feb. 26, 1995.

6. Office of Technology Assessment, *Technology against Terrorism: The Federal Effort* (Washington: GPO, 1991), 51–52.

7. Raymond Zilinskas, "Terrorists and BW: Inevitable Alliance?" *Perspectives in Biology and Medicine* 34 (Autumn 1990). Eitzen, prepared statement. Interview with Dr. Kathleen Bailey, Dec. 1995.

8. It is quite easy to purchase these manuals. I called one such publishing house and told the operator I wanted to buy manuals with instructions on how to poison people. She asked whether I was interested in bombs or silencers as well. I told her no, I only wanted to poison people. She asked for my credit card number and mailing address, and that was the end of our conversation.

9. The most prominent example is a case in which a killer followed instructions

provided in a manual entitled *Hit Man.* See *Rice v. Paladin Enterprises, Inc.*, 940 F. Supp. 836 (D. Md. 1996), appeal docketed, no. 96–2412 (4th circuit).

10. Leonard A. Cole, *Clouds of Secrecy: The Army's Germ Warfare Tests over Populated Areas* (Totowa, N.J.: Rowman and Littlefield, 1988), 163. The Army carried out the tests in the belief that the simulants were harmless to human health, but a few people were adversely affected and one hospital patient died. The Army ended such simulated attacks against human beings in the 1960s.

11. Ibid., 68. *U.S. Army Activities in the United States Biological Warfare Programs, 1942–1977* (Washington: Department of the Army, 1977), vol. 1, 6–3; vol. 2, IV-E-1-1.

12. *U.S. Army Activities in Biological Warfare Programs,* vol. 2, IV-E-5-1–5A-1.

13. See Ron Purver, "Chemical and Biological Terrorism: The Threat According to the Open Literature," Canadian Security Intelligence Service, June 1995.

14. William Patrick, "Biological Terrorism and Aerosol Dissemination," *Politics and the Life Sciences,* Sept. 1996, 209; and interviews at Fort Detrick. Eitzen, prepared statement, 112.

15. Stephen C. Reynolds, "The Terrorist Threat to Domestic Water Supplies" (U.S. Army Corps of Engineers, Aug. 1987), unclassified/limited distribution. Seymour S. Block, *Disinfection, Sterilization, and Preservation* (Philadelphia: Lea and Febiger, 1991). U.S. Army Environmental Hygiene Agency, Aberdeen Proving Ground, "Position Paper: Threat of Chemical Agents in Field Drinking Water," March 1982. Cryptosporidium lives in human and animal intestines and is secreted in feces. It can be found in most surface water, particularly after heavy rains. It is often difficult to identify and is not always killed by routine chlorination.

16. B. J. Berkowitz et al., "Superviolence: The Civil Threat of Mass Destruction Weapons" (Washington: Advanced Concept Research, 1972). Zilinskas, "Terrorists and BW."

17. Reynolds, "Terrorist Threat to Water Supplies."

18. General Accounting Office, "Commercial Nuclear Fuel Facilities Need Better Security," May 2, 1977, ii, cited in J. K. Campbell, "The Threat of Non-State Proliferation" (manuscript, Defense Intelligence Agency).

19. Letter from V. Bush to General Groves, Nov. 15, 1943, Bush-Conant Files, RG 227, folder 157, National Archives.

20. "Military Use of Radio-Active Materials and Organization for Defense," S-1 files, and "Report of the Subcommittee of the S-1 Committee on the Use of Radioactive Material as a Military Weapon," 7, in Bush-Conant Files, RG 227, folder 157, National Archives. "Fission aerial bombs or fission projectiles" may refer to atomic weapons wrapped in cobalt or another gamma emitter or, more likely, a conventional bomb containing radioactive material.

21. "Military Use of Radio-Active Materials," Bush-Conant Files.

22. Information provided by Doug Stephens, Lawrence Livermore National Laboratory. W. M. Place, F. C. Cobb, and C. G. Defferding, "Palomares Summary

Report," Field Command, Defense Nuclear Agency, Technology and Analysis Directorate, Kirtland Air Force Base (Jan. 1975). Randy Maydew, "Find the Missing H-Bomb," *Air Combat,* Nov.–Dec. 1996.

23. "Military Use of Radio-Active Materials," Bush-Conant Files.

24. David Albright, personal communication, Sept. 24, 1996.

25. Experts became concerned about the possibility of truck bombs at power plants after a car crashed through the gate at Three Mile Island in 1993. Matthew Wald, "US Examining Ways to Protect Nuclear Plants against Terrorists," *New York Times,* April 23, 1993. Sandia National Laboratory concluded in 1984 that "unacceptable damage to vital reactor systems could occur from a relatively small charge at close distances and also from larger but still reasonable size charges at large setback distances." "Weekly Information Report to the NRC Commissioners," April 20, 1984, enclosure E, 3. In 1994 the Nuclear Regulatory Commission demanded security upgrades at commercial reactors that would make a truck bomb attack more difficult.

26. Germany's federal counter-sabotage agency, the BFV, determined that escalating violence accompanying the transport of spent fuel was the "work of terrorists" who might have contacts with the Red Army Faction. "State, GNS Differ over Strategy to Defy Saboteurs at Gorleben," *Nuclear Fuel,* Jan. 13, 1997, 11.

27. Rocky Flats held 11.9 metric tons of weapons-grade plutonium as of Sept. 1994 and 2.8 metric tons of HEU as of Feb. 1996. U.S. Department of Energy, "Storage and Disposition of Weapons-Usable Fissile Materials Final Programmatic Environmental Impact Statement, Summary," DOE/EIS-0229, Dec. 1996, S-5. Jim Carrier, "Flats Security Lax, Ex-Officials Warn," *Denver Post,* May 20, 1997. James Brooke, "Plutonium Stockpile Fosters Fears of a Disaster Waiting to Happen," *New York Times,* Dec. 11, 1996. John J. Fialka, "Energy Department Report Faults Security at Weapons Plants," *Wall Street Journal,* June 16, 1997.

28. Paul Leventhal and Yonah Alexander, *Preventing Nuclear Terrorism* (Lexington, Mass.: Lexington Books, 1987), 9, 58.

29. Interview with Ambassador Thomas Graham, Nov. 25, 1996.

30. David Albright, "South Africa's Secret Nuclear Weapons," ISIS Report (Institute for Science and International Security), May 1994. Graham interview. Waldo Stumpf, "South Africa's Nuclear Weapons Programme," in Kathleen C. Bailey, *Weapons of Mass Destruction: Costs Versus Benefits* (New Delhi: Manohar, 1994), 71. Albright claims there were seven weapons, but Stumpf says only six were completed.

31. Stumpf, "South Africa's Nuclear Weapons Programme," 75, 76. Cost figures assume 1994 exchange rates (presumably in 1994 dollars).

32. Albright, "South Africa's Secret Nuclear Weapons."

33. Ibid. I was able to confirm in general terms these approximate figures in interviews with U.S. government officials who had interviewed South African nuclear specialists.

34. William Arkin of Greenpeace says he came close to buying a nuclear warhead from a Russian soldier working at a storage site in East Germany in the early 1990s. The soldier told Arkin he had found a way to gain access to the warheads during the transition between shifts. "The orientation of security," Arkin explains, "was very heavily weighted toward defending against a NATO attack. It was not heavily weighted toward protesters, or public intervention, or terrorists." Quoted in William Burrows and Robert Windrem, *Critical Mass* (New York: Simon and Schuster, 1994), 249. The CIA has noted Aum Shinrikiyo's apparent interest in buying nuclear warheads from Russia: Statement for the record by John Deutch, U.S. Senate, *Hearings on Global Proliferation of WMD,* pt. 2, S.Hrg. 104–422, 104th Cong., 2nd sess., March 20, 1996, 7. And Rensselaer Lee cites a 1991 letter reportedly faxed to the Russian nuclear weapons laboratory Arzamas-16, allegedly from Islamic Jihad, offering to buy a nuclear warhead. The director of Arzamas reportedly also told Lee that Iraqi agents had offered $2 billion for a warhead in 1993. Interview with Lee, Nov. 12, 1996. I have found no other reports of this letter.

35. Bill Gertz, "Russian Renegades Pose Nuke Danger," *Washington Times,* Oct. 22, 1996. Interview with an expert on Russian nuclear weapons, Oct. 16, 1996.

36. Unclassified cable, Moscow 13851, TOR: 0316032, June 1997.

37. This section is based on David E. Kaplan and Andrew Marshall, *The Cult at the End of the World* (New York: Crown, 1996).

38. Ibid., 28.

39. "Prosecutors Investigate Lobov's Links to Religious Sect," FBIS-SOV-97144, May 24, 1997; "Lobov Faces Questions over Investigation of Japanese Sect," FBIS-SV-97-084, April 25, 1997; both trans. from Moscow Interfax.

40. CIA, "The Chemical, Biological and Radiological Terrorist Threat from Non-State Actors," paper presented to Aspen Strategy group conference "The Proliferation Threat of Weapons of Mass Destruction and U.S. Security Interest," Aspen, Colo., Aug. 1996. William Broad, "How Japan Germ Terror Alerted World," *New York Times,* May 26, 1998.

41. Ibid. Another biological attack also failed: on March 15, 1995, cult members planned to release botulinal toxin at Kasumigaseki station, but a member struck by a guilty conscience neglected to arm the devices.

42. John F. Quinn, "Terrorism Comes to Tokyo: The Aum Shinri Kyo Incident," paper presented to the Association of Former Intelligence Officers, 1996 Annual Convention, Falls Church, Va., Oct. 1996. Authorities assumed the sarin had formed spontaneously from pesticide residues. Leonard Cole, *The Eleventh Plague* (New York: Freeman, 1996), citing *Mainichi Daily News,* June 30, July 2, July 9, and July 16, 1994.

43. See Jonathan B. Tucker, "Chemical/Biological Terrorism: Coping with a New Threat," *Politics and the Life Sciences* 15 (Sept. 1996). Ron Purver, "The Threat of Chemical/Biological Terrorism," *Commentary* 60 (Aug. 1995). Kaplan and Mar-

shall, *Cult at the End of the World.* Interviews with John Sopko. Quinn, "Terrorism Comes to Tokyo." CIA, "Chemical, Biological and Radiological Threat."

44. "Tokyo Police Find Bottle of a Cult's Deadly Gas," Associated Press, *New York Times,* Dec. 12, 1996, 15. Nicholas D. Kristof, "Tokyo Suspect in Gas Attack Erupts in Court," *New York Times,* Nov. 8, 1996, 14. Cole, *Eleventh Plague,* 155. Ed Evanhoe says that after Asahara was arrested North Korea may have moved its nuclear smuggling base of operations to Tumen, China, making use of a North Korean organized-crime ring to smuggle nuclear-related equipment as well as nuclear materials. Email from Evanhoe, Nov. 5, 1996.

45. Staff Report, U.S. Senate, *Hearings on Global Proliferation of WMD,* Nov. 1, 1995. Statement for the Record by John Deutch, 7. Nicholas D. Kristof, "Japanese Cult Said to Have Planned Nerve-Gas Attacks in U.S.," *New York Times,* March 23, 1997.

46. Staff Report, U.S. Senate, *Hearings on Global Proliferation of WMD,* Nov. 1, 1995.

47. Joseph Pilat, "World Watch: Striking Back at Urban Terrorism," *NBC Defense and Technology International* (June 1986), 18. "Animal Rights Activists Attack Scientists' Homes," Associated Press, *Los Angeles Times,* March 13, 1985. Rand database. *Product Tampering and the Threat to Tamper* (Los Angeles: Foundation for American Communications, undated), 3.

48. James Campbell, "Weapons of Mass Destruction and Terrorism: Proliferation by Non-State Actors" (Master's thesis, Naval Postgraduate School, 1996). Rand database. *Arkansas Gazette,* April 27, 1987, cited in Bruce Hoffman, " 'Holy Terror': The Implications of Terrorism Motivated by Religious Imperative," Rand Paper P-7834, 1993.

49. "New Charges Filed against Alleged Leader of Bombing," *Washington Post,* Oct. 7, 1995, 14.

50. Cesium-137, a radioisotope used in the treatment of cancer, is a waste product of nuclear reactors. It has a relatively long half-life, and areas contaminated with it require extensive cleanup. It can be absorbed into the food chain and is carcinogenic. Mark Hibbs, "Chechen Separatists Take Credit for Moscow Cesium-137," *Nuclear Fuel* 20, no. 25 (Dec. 4, 1995), 5.

51. John McQuiston, "Plot against L.I. Leaders Is Tied to Fear of UFO's," *New York Times,* June 22, 1996. "Two Charged in Plot to Poison Long Island GOP Officials; Radium, Weapons Cache Found in House," *Washington Post,* June 14, 1996. Larry Sutton, "Bail Denied in Radium Plot," *New York Daily News,* June 25, 1996.

52. Prepared Statement of John O'Neill, U.S. Senate, *Hearings on Global Proliferation of WMD,* Nov. 1, 1995, 241. Thomas J. Torok et al., "A Large Community Outbreak of Salmonellosis Caused by International Contamination of Restaurant Salad Bars," *JAMA* 278, no. 5 (Aug. 6, 1987); Seth Carus, "The Rajneesh in Oregon," paper presented at a workshop on Patterns of Behavior Associated with

Chemical and Biological Terrorism, Monterey Institute, Washington, June 1998. The Rajneeshees' true motivations are not clear. If their goal was exclusively to make their victims ill, rather than to affect a target audience (by, for example, frightening the local residents), this incident may not, strictly speaking, fit the definition of terrorism used in this book.

53. Broad, "How Japan Germ Terror Alerted World."

5. Who Are the Terrorists?

1. The IRA, for example, would probably be less effective in fundraising operations on Boston Common if it employed bubonic plague as a weapon. But recent evidence suggests that even the IRA may have been in the market for radiological materials; the British diplomats expelled from Russia in May 1996 were reportedly investigating IRA weapons transactions there. "IRA Linked to Radioactive Material," *Washington Times,* May 13, 1996.

2. Konrad Kellen, *Terrorists—What Are They Like? How Some Terrorists Describe Their World and Actions* (Santa Monica: Rand, 1979), 62.

3. Hans Zinsser, *Rats, Lice and History* (1933; Boston: Little, Brown, 1963), 110.

4. The Phineas story was also the core myth for the Zealots-Sicarii, described in Chapter 2. David C. Rapoport, "Fear and Trembling: Terrorism in Three Religious Traditions," *American Political Science Review* 78, no. 3 (Sept. 1984), 669.

5. Not all millenarian doctrines are religiously based. Nineteenth-century Russian anarchists used the millenarian ideas to justify acts of violence. Secular millenarians envision a new world created by human beings; religious millenarians envision a divine power who will destroy the corrupt world and create a new one cleansed of the infidels. Jean E. Rosenfeld, "Pai Marire: Peace and Violence in a New Zealand Millenarian Tradition," *Terrorism and Political Violence* 7, no. 3 (Autumn 1995), 83.

6. Interview with Kerry Noble, March 2, 1998.

7. Rosenfeld, "Pai Marire," 101–102.

8. " 'Millennial Madness' Said to Drive Fringe Groups," *Washington Times,* March 29, 1997, 3.

9. "Millennial Prophecy Report," *Guardian,* April 8, 1995. Walter Lacquer, "Fin-de-siecle: Once More with Feeling," *Journal of Contemporary History* 31, no. 1 (Jan. 1996), 39, 40.

10. David E. Kaplan and Andrew Marshall, *The Cult at the End of the World* (New York: Crown, 1996), 66.

11. Interview with William Pierce, April 22, 1997.

12. Staff Statement, U.S. Congress, Senate, Committee on Governmental Affairs, Permanent Subcommittee on Investigations, *Hearings on Global Prolifer-*

ation of Weapons of Mass Destruction, 104th Cong., 1st sess., pt. 1, Oct. 31, 1995.

13. I thank Martha Crenshaw for this insight. See also Brian Jenkins, "The Constraints of Terror," *Harvard International Review* 17, no. 3 (1995).

14. U.S. Department of State, *Patterns of Global Terrorism 1989* (Washington, 1990), 51–52. Jeffrey Simon, *Terrorists and the Potential Use of Biological Weapons* (Santa Monica: Rand, 1989), 13.

15. Jack Anderson, "Chemical Arms in Terrorism Feared by CIA," *Washington Post,* Aug. 27, 1984, C-14. Brian Jenkins, remarks at a conference on ChemBio Terrorism, Washington, April 29, 1996.

16. *Patterns of Global Terrorism 1989,* 11.

17. Jenkins has often made this statement about what terrorists want: e.g., Brian Michael Jenkins, "International Terrorism: A New Mode of Conflict," in David Carlton and Carolo Schaerf, eds., *International Terrorism and World Security* (London: Croom Helm, 1975), 15. In speeches since the Aum Shinrikiyo attack he has noted that some terrorists do seem to want a lot of people dead. Kenneth Waltz, "Waltz Responds to Sagan," in Scott D. Sagan and Kenneth Waltz, *The Spread of Nuclear Weapons: A Debate* (New York: Norton, 1995), 94–96.

18. Ramzi Ahmed Youssef, the mastermind of the bombing, told an FBI agent that the World Trade Center towers "wouldn't be [standing] if I had had enough money and explosives." Benjamin Weiser, "Trial Begins for Chief Suspect in Trade Center Blast," *New York Times,* Aug. 4, 1997, 16. *Terrorism in the United States* (Washington: U.S. Department of Justice, 1993). "Sentencing Statement of Judge Duffy, Trial Judge of World Trade Bombing Case," in appendix to U.S. Senate, *Hearings on Global Proliferation of WMD,* pt. 3, 104th Cong., 2nd sess., pt. 2, March 27, 1996, 276. Six people died in the bombing, and more than 1,000 were injured. The incident was very costly for insurers. "World Trade Center Insurers Have Paid $510 Million So Far," *Washington Post,* March 30, 1993, 12. Not all engineers accept the FBI's assessment of the imminence of collapse.

19. Robert D. McFadden, "FBI Seizes 8, Citing a Plot to Bomb New York Targets and Kill Political Figures," *New York Times,* June 25, 1993, 1. Jim McGee and Lynne Duke, "Informer Played Key Role in NY Bomb Plot Arrests," *Washington Post,* June 27, 1993, 1.

20. B. J. Berkowitz describes six psychological types who would be most likely to threaten or try to use WMD: paranoids, paranoid schizophrenics, borderline mental defectives, schizophrenic types, passive-aggressive personality types, and sociopath personalities. He considers sociopaths the most likely actually to use WMD. Berkowitz et al., *Superviolence: The Civil Threat of Mass Destruction Weapons* (Santa Barbara: ADCON Corp., 1972), 3-9, 4-4.

21. Ted Robert Gurr, "Terrorism in Democracies," in Walter Reich, ed., *Origins of Terrorism: Psychologies, Ideologies, Theologies, States of Mind* (New York: Cambridge University Press, 1990), 102.

22. Thomas Schelling, "What Purposes Can 'International Terrorism' Serve?"

in R. G. Frey and Christopher W. Morris, eds., *Violence, Terrorism, and Justice* (New York: Cambridge University Press, 1991), 20–21.

23. Pierce interview.

24. Ibid. Andrew Macdonald [pseudonym for William Pierce], *The Turner Diaries* (New York: Barricade Books, 1996).

25. Albert Bandura, "Mechanisms of Moral Disengagement," in Reich, ed., *Origins of Terrorism.*

26. The Bruder Schweigen's Declaration of War, published on the World Wide Web.

27. Donatella della Porta, "Political Socialization in Left-Wing Underground Organizations: Biographies of Italian and German Militants," in della Porta, ed., *Social Movements and Violence: Participation in Underground Organizations,* vol. 4 (Greenwich, Conn.: JAI Press, 1992), 286. Maria Moyano, "Going Underground in Argentina: A Look at the Founders of a Guerrilla Movement," ibid.

28. Bandura, "Mechanisms of Moral Disengagement," 176. Della Porta, "Political Socialization."

29. Nearly identical stories on biological genocide were printed in *The Jubilee,* Aryan Nations' *Calling Our Nation,* Militia of Montana's *Taking Aim, The Spotlight, The American's Bulletin, Anti-Shyster,* and *The Free American.* Similar reports were heard on short-wave broadcasts. *Klanwatch Intelligence Report,* Winter 1997, 9.

30. Lieutenant Colonel Dave Grossman, *On Killing: The Psychological Cost of Learning to Kill in War and Society* (Boston: Little, Brown, 1995), 119.

31. Aryan Nations website.

32. Grossman, *On Killing,* 250.

33. J. L. Sleeman, *Thugs; or, A Million Murders* (London: S. Low and Marston, 1933), 3–4, cited in Rapoport, "Fear and Trembling," 664.

34. L. Boellinger, "Die Entwicklung zu terroristischem Handeln als Psychosozialer Prozess," in H. Jaeger, G. Schmidtchen, and L. Suellwold, eds., *Lebenslaufanalysen* (Oplader: Westdeutscher, 1981), 203, quoted in della Porta, "Political Socialization."

35. I. Janis, *Victims of Groupthink* (Boston: Houghton Mifflin, 1972); Jerrold Post, "Terrorist Psycho-Logic," in Reich, ed., *Origins of Terrorism.*

36. The guerrillas were partly responsible for the subsequent breakdown in democracy, Moyano argues, and the establishment of a highly repressive regime. Moyano, "Going Underground." Donatella della Porta, "Research on Individual Motivations in Underground Political Organizations," in della Porta, *Social Movements and Violence.*

37. Kellen, *Other Terrorists,* 62.

38. "Leaderless Resistance, An Essay by L. R. Beam," published on the World Wide Web. This system of organization, Beam claims, is almost identical to "the methods used by the committees of correspondence during the Ameri-

can Revolution." It is also similar in structure to Communist revolutionaries' cells.

39. Ariel Merari, "The Readiness to Kill and Die: Suicidal Terrorism in the Middle East," in Reich, ed., *Origins of Terrorism,* 205. Martin Kramer, "The Moral Logic of Hizballah," ibid., 142n24.

40. Kramer, "Moral Logic of Hizballah," 144–146.

41. Merari, "Readiness to Kill and Die," 205. Thomas Friedman, "Boy Says Lebanese Recruited Him as a Car Bomber," *New York Times,* April 14, 1985, 1; cited in Walter Reich, "Understanding Terrorist Behavior," in Reich, ed., *Origins of Terrorism,* 271. Kramer, "Moral Logic of Hizballah," 143.

6. The Threat of Loose Nukes

1. Tim Zimmerman, "The Russian Connection," *U.S. News & World Report* 119 (23 Oct. 1996).

2. The facility, which is involved in designing breeder reactors, was also the source of smuggled plutonium in an apparently unrelated incident. Doug Clarke and Penny Morvant, "FSB: Smuggled Plutonium Was from Russia," *OMRI Daily Digest,* Feb. 13, 1996. Much of the material in this chapter is based on interviews in Moscow between Nov. 3 and Nov. 15, 1995, with officials from the Ministry of Atomic Energy (Minatom), Gosatomnadzor (Russia's nuclear regulatory agency), and the Security Council, who asked not to be identified by name.

3. Organized crime has allegedly been involved in other cases of nuclear-related trafficking. E.g., see Nikolay Lashkevich, "Nuclear Mafia Goes out onto Market," *Izvestiya,* Jan. 31, 1995.

4. Sarah A. Mullen et al., *The Nuclear Black Market* (Washington: Center for Science and International Studies, 1996), 5.

5. Vladimir Orlov, comments delivered at the Carnegie Endowment for International Peace Conference, "Nuclear Non-Proliferation: Enhancing the Tools of the Trade," Washington, June 10, 1997.

6. Senior U.S. intelligence official, off-the-record talk, Washington, June 2, 1997.

7. Penny Morvant, "Army Housing Shortage Continues," *OMRI Daily Digest* 2 (April 30, 1996).

8. Scott Parrish, "Chief of General Staff: Officer Corps Decaying," *OMRI Daily Digest* 3 (March 17, 1997). Penny Morvant, "Crimes Increasing in Military," *OMRI Daily Digest* 2 (Oct. 16, 1996). Interview with CIA analyst, June 19, 1997. "Prosecutors Investigating 21 Generals for Corruption," *RFE/RL Newsline,* Nov. 5, 1997. Penny Morvant, "Illegal Arms Trade," *OMRI Daily Digest* 2 (Oct. 9, 1996). Scott Parrish, "Deputy Warns Military Is in 'Explosive' Situation," *OMRI Daily Digest* 2 (Aug. 20, 1996). Richard Beeston, "Russian Military Warns of Mutiny over Troop Cuts," *Times* (London), July 1, 1997. Deborah Yarsike Ball, "How Reliable Are

Russia's Officers?" *Jane's Intelligence Review* 8, no.5 (May 1996), 204. Scott Parrish, "Lebed Visits Strategic Missile Forces Command," *OMRI Daily Digest* 2 (Oct. 14, 1996).

9. Testimony by David Osias, U.S. Congress, Senate, Foreign Relations Committee, *Loose Nukes, Nuclear Smuggling, and the Fissile Materials Problem in Russia and the NIS: Hearings before the Subcommittee on European Affairs,* 104th Cong., 1st sess., Aug. 22, 1995. Bill Gertz, "Russian Renegades Pose Nuclear Danger," *Washington Times,* Oct. 22, 1996.

10. Osias testimony.

11. Unclassified Cable Moscow 13851, TOR: 031603Z, June 1997. Interviews with House National Security Committee staff, June 5 and 10, 1997. Lebed has repeatedly changed his story on both the number and type of weapons alleged to be missing. The Russian government first denied that such small nuclear weapons had ever been made; later it seemed to concede their existence, and even that their security might be imperfect. Ivo Dawnay, "Russia Loses Its Suitcase N-bombs," *Electronic Telegraph* (U.K.), Nov. 9, 1997. "Russia Does Have Suitcase Bombs, Lebed Says," *BBC Summary of World Broadcasts,* Nov. 22, 1997.

12. House National Security Committee staff interviews. Lebed wants to undermine the Yeltsin government. Other officials may have similar incentives, or may want to scare the United States into supplying more funding for nuclear security.

13. Indictment against Alexander Darichev and Aleksandr Pogrebezskij, U.S. District Court, Southern District of Florida, Case no. 97–0541; Affidavit of S/A William P. Eshleman in Support of Complaints and Warrants for Arrest. "Two Lithuanians Arrested in Miami, Are Accused of Bid to Sell Soviet Nuclear Weapons," *New York Times,* July 1, 1997. "Lithuanian Nuclear Smugglers 'Swindlers,' " Reuters Newswire, July 2, 1997.

14. Americans use the term "football" for an electronic device that allows the President to authorize the use of nuclear weapons. It is a command and communications link with the Department of Defense (in Russia, the Ministry of Defense). In Russia, President Yeltsin, the Minister of Defense, and the Chief of General Staff have attaché-sized "footballs," which together they would use both for receiving warnings about incoming missiles and for transmitting the command to launch nuclear weapons. The CIA reportedly says the device is largely symbolic and may not be capable either of activating a strike or of blocking the general staff from using nuclear weapons on their own initiative. General Yevgeny Shapashnikov, President Yeltsin's first Defense Minister, was once asked if he and Yeltsin together could "push the button" launching a nuclear strike, and reportedly replied, "Yes, but nothing would happen." Bill Gertz, "Russia's Nuclear 'Football' Easy to Block," *Washington Times,* Oct. 22, 1996, 6.

15. Bruce Blair, "Russian Nuclear Control: Risks and Solutions," paper presented in Helsinki, Aug. 26, 1997. Peter Pry, *War Scare: Nuclear Countdown after*

the Soviet Fall (Atlanta: Turner Publications, forthcoming). Scott Parrish, "Yeltsin, Rodionov Discuss Nuclear Arsenal," *OMRI Daily Digest* 3 (March 18, 1997).

16. Gertz, "Russian Renegades Pose Nuclear Danger." Scott Parrish, "CIA: Russian Nuclear Safeguards Weakening," *OMRI Daily Digest* 2 (Oct. 23, 1996). Scott Parrish, "Russia Says Nuclear Arms under Control," ibid. (Oct. 24, 1996).

17. E.g., John Stewart, former director of DOE's Office of Foreign Intelligence, believes that "the disintegration of Russian command and control is the most serious security threat facing the United States."

18. The figure 600 storage sites is from testimony by Gordon Oehler, U.S. Congress, Senate, Armed Services Committee, *Intelligence Briefing on Smuggling of Nuclear Material and the Role of International Crime Organizations, and on the Proliferation of Cruise and Ballistic Missiles,* S. Hrg. 104–35, 104th Cong, 1st sess., Jan. 31, 1995, 4. General Maslin said there were 200 sites in 1991. Maslin quoted in Vladimir Orlov, "A Threat of Nuclear Terrorism Exists in Russia," *Moskovskiye Novosti* no. 44, June 25, 1995. Another official described the vulnerability of warheads in transit at a parliamentary hearing, claiming that warheads are still not protected against grenade throwers and bullets. "We Cannot Preclude the Possibility of Nuclear Theft," *Yaderny Kontrol* 5 (Fall 1997), 12.

19. Interview with Minatom official, Nov. 4, 1995. A DOE official points out that a simple procedural change could facilitate verification of nuclear material.

20. The Soviet Union's declared defense budget excluded items like military R&D, stockpiling, and civil defense, and pricing practices were quite different from those in the West. *The Military Balance 1984–1985* (London: International Institute for Strategic Studies, 1984), 15. *The Military Balance 1986–1987* (London: International Institute for Strategic Studies, 1986), 32–33.

21. Penny Morvant, "Nuclear Specialists Protest Wage Delays," *OMRI Daily Digest* 2 (June 11, 1996). Official unemployment rates have climbed as high as 12% in Penza, for example, but unofficial estimates are as high as 20%.

22. The official name of Arzamas-16 is the All-Russian Scientific Research Institute of Experimental Physics (VNIIEF) at Sarov. Alan Cooperman and Kyrill Belianinov, "Moonlighting by Modem in Russia," *U.S. News & World Report,* April 17, 1995. "Press Conference with Nuclear Energy Minister Viktor Mikhailov," Official Kremlin International News Broadcast, Feb. 18, 1998. "Russia: Ministries to Draft Decision on Nuclear Centers," FBIS-SOV-97–156, June 5, 1997; source: LD0506122797 Moscow Interfax in English 1130 GMT June 5, 1997. "Official Claims It Is Possible and Easy to Steal a Nuclear Bomb," *BBC Summary of World Broadcasts,* July 16, 1997.

23. Dorothy S. Zinberg, *The Missing Link? Nuclear Proliferation and the International Mobility of Russian Nuclear Experts* (Geneva: United Nations Institute for Disarmament Research, 1995), 3. Cooperman and Belianinov, "Moonlighting by Modem." Jim Hoagland, "Hammering at Russia," *Washington Post,* Jan. 8, 1988, A21.

24. "Desperate Nuclear Scientist Commits Suicide," Reuters, Oct. 31, 1996. The official name of Chelyabinsk-70 is the All-Russian Scientific Research Institute of Technical Physics (VNIITF) at Snezhinsk.

25. Statement for the record by John Deutch, U.S. Congress, Senate, Committee on Governmental Affairs, Permanent Subcommittee on Investigations, *Hearings on Global Proliferation of Weapons of Mass Destruction,* pt. 2, S.Hrg. 104–422, 104th Cong., 2nd sess., March 20, 1996, 3. "Russia Plans to Double Nuclear Exports," *RFE/RL Newsline* 1, no. 59 (June 24, 1997). "Press Conference with Nuclear Energy Minister Viktor Mikhailov," Official Kremlin International News Broadcast, Feb. 18, 1998.

26. "Russia Plans to Double Nuclear Exports." Russia defends the sale of the reactor as legal under the NPT, and as no more risky from the point of view of nuclear proliferation than the reactor now promised to North Korea, to be provided by a consortium of countries, including South Korea, the United States, and Russia. The U.S. government claims the two sales are quite different: North Korea is already experienced with reactor technology, and the reactor provided by the international consortium is proliferation-resistant. Moreover, Iran's objectives are suspect: why would a country that is flaring natural gas (because it has a surplus too expensive to export) need a powerful, expensive nuclear reactor it cannot afford?

27. Alexei Yablokov, "Atomic Energy Ministry Mixed Up Its Own Interests with National Ones," *Izvestiya,* June 2, 1995, 3. "Atomic Energy Minister Threatens West over NATO Expansion," *OMRI Daily Digest* 2 (Feb. 16, 1996).

28. Interview with Minatom official, Nov. 4, 1995. Mark Hibbs, "Physical Protection Reportedly Eroding at Minatom's 10 'Closed Cities' in Russia," *Nuclear Fuel* 20, no. 1 (Jan. 2, 1995), 13.

29. General Accounting Office, *Weapons of Mass Destruction: Reducing the Threat from the Former Soviet Union: An Update,* Letter Report (Appendix 3), GAO/NSIAD-95–165 (Washington: GPO, June 9, 1995). Gennadi Yezhov, "Chernomyrdin Does Not Preclude Chance of Nuclear Thefts," *Itar-Tass,* Feb. 23, 1995. ". . . Amid Concern over Army's Role," *Jamestown Foundation Monitor,* Jan. 22, 1996. "Official Claims It Is Possible and Easy to Steal a Nuclear Bomb."

30. U.S. Department of Energy, "Nuclear Sites of Russia and the Newly Independent States of the Former Soviet Union" (Richland, Wash: Pacific Northwest Laboratory, 1995), 50.

31. The U.S. wanted Russia to take the Tbilisi material, but the Russian government was reluctant to do so. Michael Gordon, "Russia Thwarting U.S. Bid to Secure a Nuclear Cache," *New York Times,* Jan. 5, 1997, 1. Misha Dzhindzhikhashvili, "Georgia Offers Uranium for Sale," Associated Press, Jan. 7, 1997. U.S. Dept. of Energy, "Nuclear Sites." Interview with an American familiar with the Caucasus region, June 3, 1997.

32. Interview with a senior member of President Yeltsin's Security Council staff, Nov. 8, 1995.

33. Hibbs, "Physical Protection Reportedly Eroding," 13.

34. William Potter, "Before the Deluge? Assessing the Threat of Nuclear Leakage from the Post-Soviet States," *Arms Control Today,* Oct. 1995. Rensselaer Lee, "Smuggling Update," *Bulletin of the Atomic Scientists* 53, no. 2 (May/June 1997), 55. The German government announced that the Russian Federal Security Service (FSB) had sent a letter admitting that the 10.5 ounces of Pu-239 seized in Munich in 1994 had originated in Obninsk, but the Russian government did not publicly confirm that it had sent the letter. Clarke and Morvant, "FSB: Smuggled Plutonium Was from Russia."

35. James Wyllie, "Iran—Quest for Security and Influence," *Jane's Intelligence Review* 5, no. 7 (July 1, 1993), 311. Mark Hibbs, "Kazakhs Say Iran Sought LEU for VVER Fuel, Not 'Sapphire' HEU," *Nuclear Fuel,* July 17, 1995, 11–12.

36. *Ankara Anatolia,* in English, 10:15 GMT, Oct. 6, 1993, as reported in an unclassified cable, serial TA06 10103193. *Istanbul Turkiye,* in Turkish, Oct. 7, 1993, as reported in an unclassified cable, serial NC0910082 693. Conversation with Ozgen Acar, a reporter for *Cumhuriyet,* a major Turkish daily newspaper.

37. "Iran, Iraq Secretly Buying on Nuclear Market," Reuters World Service, Jan. 18, 1996. "Iran Objects over German Nuclear Charges," Reuters World Service, Jan. 25, 1996. CTK National News Wire, May 7, 1996.

38. Interview with Minatom official, Nov. 4, 1995.

39. James Adams cites Russians claiming that their government produced some 300,000 tons of chemical agent. Adams, *The New Spies* (London: Pimlico, 1995), 279.

40. Vil S. Mirzayanov, "Dismantling the Soviet/Russian Chemical Weapons Complex: An Insider's View," in Amy Smithson et al., *Chemical Weapons Disarmament in Russia: Problems and Prospects* (Washington: Henry L. Stimson Center, 1995), 29. Prepared Statement by Vil Mirzayanov, U.S. Senate, *Hearings on Global Proliferation of WMD,* 104th Cong., 1st sess., pt. 1, Nov. 1, 1995.

41. Mark Urban, "Is Nerve Gas Russia's Next Nightmare?" *Sunday Telegraph,* Jan. 21, 1996, 21. Leonid Berres and Aleksandr Koretsky, "The Russian Federation: Crime," *Kommersant-Daily,* Oct. 24, 1995, rpt. in *Current Digest of the Post-Soviet Press,* Nov. 22, 1995.

42. "Keystone Kops on Job in Russia," *Chicago Tribune,* Oct. 26, 1995.

43. *Komsomol'skaya Pravda,* May 27, 1992. Christopher Andrew, "Coping with the Changing Face of Terror: From the Bomb and the Bullet to Nerve Gas and Bubonic Plague," *Newsday,* April 23, 1995. John Barry, "Planning a Plague?" *Newsweek,* Feb. 1, 1993, 40–41. Ian Brodie, "Russia Fails to End Its Germ War Research" *Times* (London), April 9, 1994. R. Jeffrey Smith, "U.S. Officials Allege That Russians Are Working on Biological Arms," *Washington Post,* April 8, 1994. Adams, *The New Spies.* "Russians Develop New Deadly Anthrax Strain," Reuters

Newswire, Feb. 13, 1998. Speech by James Adams, Carnegie Endowment for International Peace, Nov. 13, 1997. Adams refused to reveal his sources other than to say they were inside the CIA.

44. "Russian Security Inadequate for Chemical Weapons Storage," Agence France Presse, Aug. 2, 1995. "Russia Lacks Funds to Destroy Chemical Weapons," RFE/RL Newsline, June 23, 1997.

45. Interviews with U.S. government officials who asked not to be identified, May 26 and 27, 1997.

46. Ibid.

47. J. Terry Conway, "Project Amber," *Customs Today* 31, no. 1 (Winter 1996), 10.

48. Robert Sharlet, "Russian Constitutional Crisis: Law and Politics under Yeltsin," *Post-Soviet Affairs* 9, no. 4 (Oct.–Dec. 1993), 314.

49. Interviews in Moscow, Nov. 1995.

50. "Duma Accused of Nonfeasance," *Jamestown Monitor,* June 7, 1996. In a letter to the Duma speaker, Yegor Stroyev, the speaker of the upper house of the Russian parliament, admonished the Duma: "Lawmakers should be law-abiding." He complained that "more than a third" of the bills submitted by the present Duma had had to be rejected by the upper house because they violated the constitution or failed to specify sources of funding for the measures they called for. He said that Russia still lacks essential laws on the procuracy and the judicial system and laws regulating government intervention in the economy.

51. Laura Belin, "Russia Sets Record in Journalist Deaths," *OMRI Daily Digest* 2 (Oct. 2, 1996).

52. Stephen Handelman, "Can Russia's Mafia Be Broken?" *New York Times,* Nov. 10, 1996. Jonas Bernstein, "Murder in Moscow," *Wall Street Journal,* Nov. 7, 1996, A-22. Louise Shelly, "The Price Tag of Russia's Organized Crime," report written for the World Bank, March 20, 1997. Arnaud de Borchgrave, testimony before the House Committee on International Relations, 105th Cong., 1st sess., Oct. 1, 1997. De Borchgrave reports that organized crime controls 40% of Russian business, 60% of the remaining state-owned enterprises, and more than half of Russia's 1,740 banks.

53. Testimony of Louis Freeh, U.S. Congress, House, Committee on International Relations, 105th Cong., 1st sess., Oct. 1, 1997. Douglas Farah, "FBI Chief: Russian Mafias Pose Growing Threat to U.S.," *Washington Post,* Oct. 2, 1997, 18. Douglas Farah, "Russian Mob, Drug Cartels Joining Forces," *Washington Post,* Sept. 29, 1997, 1. Bill Gertz, "Russian Smuggling Ring Arms Kurd Rebels in Turkey," *Washington Times,* June 23, 1997.

54. De Borchgrave testimony, Oct. 1, 1997. James Adams, "The Dangerous New World of Chemical and Biological Weapons," in Brad Roberts, ed., *Terrorism with Chemical and Biological Weapons* (Alexandria, Va.: Chemical and Biological Arms Control Institute, 1997).

55. Information on border guards from DOD, Nov. 27, 1995. Dave Bhavna, "Russia to Retain Customs Controls on Kazakhstani Border," *OMRI Daily Digest* 2 (Feb. 5, 1996).

7. The State as Terrorist

1. I thank Greg Koblentz and Steve Black for their assistance with this chapter: Greg was responsible for much of the research.

2. Staff Report, U.S. Congress, Senate, Foreign Relations Committee, *Chemical Weapons Used in Kurdistan: Iraq's Final Offensive,* 100th Cong., 2nd sess., Oct. 1988. "Winds of Death: Iraq's Use of Poison Gas against Its Kurdish Population," report by Physicians for Human Rights, Somerville, Mass.

3. Details of Iraq's chemical weapons efforts from, e.g., Thatcher and Aeppel, "The Trail to Samarra," *Christian Science Monitor,* Dec., 13, 1988, B1; Herbert Krosney, "Poison Gas: Iraq's Deadly Weapon of Last Resort," press release from L.A. Times Syndicate International, July 29, 1988. Iraq signed the Geneva Protocol "On condition that the Iraq Government shall be bound by the provisions of the Protocol only towards those States which have both signed and ratified it or acceded thereto, and that they shall not be bound by the Protocol towards any State at enmity with them whose armed forces or the forces of whose allies, do not respect the disposition of the Protocol." As of 1987, 112 countries had acceded to the Protocol for the Prohibition of the Use in War of Asphyxiating, Poisonous or Other Gases, and of Bacteriological Methods of Warfare (June 17, 1925); 35 with the reservation that CW could be used in retaliation. *Status of Multilateral Arms Regulation and Disarmament Agreements* (New York: United Nations, 1987).

4. Although the Australia Group founded in response to revelations about Iraq's chemical program did limit exports of key precursors, the export controls were too little punishment, imposed too late. Iraq knew how to make the chemicals by then. This was not the only time the Geneva Protocol was violated without causing significant international outcry: two examples are Italy's use of chemical weapons against the Abyssinians and Egypt's against Yemen.

5. "Britain Cites Iraqi Threat," *New York Times,* March 11, 1998, A11.

6. Jonathan Tucker, who worked as an inspector in Iraq, says aflatoxin may have incapacitating effects in high-dose aerosols. Asked what Iraq intended to do with such a weapon, David Kay claimed that the most likely use was to target Kurds. Kay, speech before the American Bar Association Committee on Law Enforcement and National Security, March 19, 1998.

7. R. Jeffrey Smith, "UN Pursuing 25 Germ Warheads It Believes Are Still Loaded with Deadly Toxin," *Washington Post,* Nov. 21, 1997, 1. Bruce Nelan, "America the Vulnerable," *Time* 150, no. 22 (Nov. 24, 1997).

8. Kenneth N. Waltz, *The Spread of Nuclear Weapons: More May Be Better,* Adelphi Paper no. 171 (London: International Institute for Strategic Studies, 1981).

Bruce Beuno de Mesquita and William H. Riker, "An Assessment of the Merits of Selective Nuclear Proliferation," *Journal of Conflict Resolution* 26, no. 2 (June 1982), 283.

9. Scott Sagan, "The Perils of Proliferation: Organization Theory, Deterrence Theory, and the Spread of Nuclear Weapons," *International Security* 18, no. 4 (Spring 1994). See Peter R. Lavoy, "The Strategic Consequences of Nuclear Proliferation," *Security Studies* 4, no. 4 (Summer 1995).

10. John P. Sutherland, "The Story General Marshall Told Me," *U.S. News and World Report* 4, no. 18 (Nov. 2, 1959), 52.

11. Scott Sagan, "Why Do States Build Nuclear Weapons? Three Models in Search of a Bomb," *International Security* 21, no. 3 (Winter 1996). Jeffrey Legro, "Which Norms Matter? Revisiting the 'Failure' of Internationalism," *International Organization* 51, no. 1 (Winter 1997). Richard Price and Nina Tannenwald, "Norms and Deterrence: The Nuclear and Chemical Weapons Taboos," in Peter Katzenstein, ed., *The Culture of National Security* (New York: Columbia University Press, 1996). Michael Mandelbaum, *The Nuclear Revolution* (New York: Cambridge University Press, 1981).

12. Sagan, "Why Do States Build Nuclear Weapons?"

13. News Conference Remarks by President Kennedy on Nuclear Tests [Extract], March 21, 1963, *Washington Post,* March 22, 1963, in Arms Control and Disarmament Agency, *Documents on Disarmament 1963* (Washington: GPO, 1964), 113.

14. David Kay, "Detecting Cheating on Non-Proliferation Regimes: Lessons from the Iraqi Experience," Aspen Strategy Group Meeting, Aug. 1996.

15. David Kay, "Denial and Deception Practices of WMD Proliferators: Iraq and Beyond," *Washington Quarterly,* Jan. 1995. Ben Sanders recalls that the agency's management "specifically excluded the possibility" of demanding special inspections at undeclared sites. Sanders, "IAEA Safeguards: A Short Historical Background," in *A New Nuclear Triad: The Non-Proliferation of Nuclear Weapons, International Verification and the International Atomic Energy Agency* (Southampton: Mountbatten Center for International Studies, 1992), 10.

16. David Fischer says the result of the IAEA's preoccupation with nuclear material is that 70% of the IAEA's safeguards budget is absorbed by Canada, Germany, and Japan. Fischer, "Innovations in IAEA Safeguards to Meet the Challenge of the 1990s" in *A New Nuclear Triad,* 28. Efforts are now under way to expand the IAEA's access to additional sites. The IAEA has the authority to go to undeclared sites in nonnuclear states, but when it announced its intention to carry out its first suspect-site inspection in North Korea, North Korea announced its intention to withdraw from the NPT.

17. INFCIRC/153, par. 42, rpt. in L. Manning Muntzing, ed., *International Instruments for Nuclear Technology Transfer* (La Grange Park, Ill.: American Nuclear Society, 1978), 145. Interview with Zachary Davis, April 1993. Larry Schein-

man, "The Current Status of IAEA Safeguards" in *A New Nuclear Triad*, 21.

18. U.S. Department of Defense, *Proliferation: Threat and Response* (Washington: GPO, April 1996), 15, 21.

19. "Iraq Foreign Minister Tariq Aziz . . . Confirms Once Again His Country's Use of CW against Iran 'from Time to Time,' but States That Iran Had Used It First, 'from the Very Beginning' of the War," *Chemical Weapons Convention Bulletin*, Autumn 1988, 6.

20. Steve Weissman and Herbert Krosney, *The Islamic Bomb* (New York: Times Books, 1981), 93–94, 98–99. Jed Snyder, "The Road to Osiraq: Baghdad's Quest for the Bomb," *Middle East Journal* 37, no. 4 (Autumn 1983), 577. Richard Burt, "U.S. Says Italy Sells Iraq Atomic Bomb Technology," *New York Times*, March 18, 1980, A1.

21. Weissman and Krosney, *Islamic Bomb*, 228, 239, 243. Shlomo Nakdimon, *First Strike* (New York: Summit, 1987).

22. Iraq continued to pursue plutonium by studying the construction of a reactor, but this effort was apparently suspended in mid-1988. IAEA, S/1996/261, April 11, 1996, 5. Iraq also negotiated with the Soviet Union, a French-Belgian consortium, and China on buying an underground reactor, but no purchase was made. Kathleen Bailey, *The UN Inspections in Iraq: Lessons for On-Site Verification* (Boulder: Westview, 1995), 71. But continued interest in a plutonium-production reactor is evidenced by continued experiments with plutonium separation and heavy water production. David Albright, Frans Berkhout, and William Walker, *Plutonium and Highly Enriched Uranium 1996: World Inventories, Capabilities and Policies* (New York: Stockholm International Peace Research Institute, 1997), 314. Iraq admitted having separated about 6 grams of plutonium in violation of its safeguards agreement with the IAEA. International Atomic Energy Agency, *IAEA Inspections and Iraq's Nuclear Capabilities*, April 1992, 7.

23. Albright, Berkhout, and Walker, *Plutonium and HEU 1996*, 314–315. Al Tarmiya was built by a foreign company, but Iraq built Ash Sharqat on its own using blueprints from the foreign firm. At the time of the Gulf War Iraq had produced only 1.3 kg of low-enriched uranium, Al Tarmiya had only 8 electromagnetic isotope-separation units installed, and the construction of the buildings at Ash Sharqat was only 85% complete. David Albright and Robert Kelley, "Has Iraq Come Clean at Last?" *Bulletin of the Atomic Scientists*, Nov.–Dec. 1995, 56. IAEA, S/22986, Aug. 28, 1991, 5.

24. IAEA, S/1995/1003, Dec. 1, 1995, 10–11. David Kay, "Interim Status Report," Aug. 12, 1991, 3. IAEA, S/1995/1003, Dec. 1, 1995, 13. Interview with UNSCOM official, March 18, 1998.

25. IAEA, S/1995/1003, Dec. 1, 1995, 3–6; S/1996/14, Jan. 10, 1996, 8; S/1996/261, April 11, 1996, 8.

26. IAEA, S/1995/1003, Dec. 1, 1995, 8. Albright, Berkhout, and Walker, *Plutonium and HEU 1996*, 345.

27. UNSCOM, S/1994/750, June 24, 1994, 18; S/1995/284, April 10, 1995, 11; S/1995/864, Oct. 11, 1995, 18; S/1996/258, April 11, 1996, 16. The White House reported that Iraq was capable of producing between 20 and 130 tons of VX. White House Fact Sheet, "Iraq's Program of Mass Destruction," Nov. 14, 1997. Interviews with UNSCOM officials.

28. UNSCOM, S/1994/1422/Add.1, Dec. 15, 1994, 8; S/23615, Oct. 25, 1991, 27.

29. UNSCOM, S/1994/750, June 24, 1994, 5; S/1996/848, Oct. 11, 1996, 21.

30. UNSCOM, S/1995/864, Oct. 11, 1995, 18, 11. Briefing by Rolf Ekeus, U.S. Congress, Senate, Committee on Governmental Affairs, Permanent Subcommittee on Investigations, *Hearings on Global Proliferation of Weapons of Mass Destruction,* pt. 2, S.Hrg. 104–422, 104th Cong., 2nd sess., March 20, 1996, 3.

31. DOD, *Proliferation: Threat and Response,* 21. Staff Report, U.S. Senate, *Chemical Weapons Use in Kurdistan,* 3. David Segal, "The Iran-Iraq War: A Military Analysis," *Foreign Affairs,* Summer 1988. Tom McNaugher, "Ballistic Missiles and Chemical Weapons: The Legacy of the Iran-Iraq War," *International Security,* Fall 1990, 6. Interview with Charles Duelfer, Feb. 20, 1997.

32. UNSCOM, S/1997/774, Oct. 6, 1997, 38. Prepared statement by Jonathan B. Tucker, U.S. Congress, House, Committee on Government Reform and Oversight Subcommittee on Human Resources, *Low-Level Chemical Weapons Exposures during the 1991 Persian Gulf War,* 105th Cong., 1st sess., April 24, 1997. Edith Lederer, "Biological Weapons," Associated Press, April 24, 1997. UNSCOM, S/1995/864, Oct. 11, 1995, 22–28.

33. UNSCOM, S/1995/864, Oct. 11, 1995, 22–28. Smith, "UN Pursuing 25 Germ Warheads," 1.

34. Interview with a U.S. government scientist who asked not to be identified. UNSCOM, S/1997/774, Oct. 6, 1997, 38. William Patrick, "Biological Terrorism and Aerosol Dissemination," *Politics and the Life Sciences,* Sept. 1996, 210. Interviews at Fort Detrick.

35. Interview with Ambassador Rolf Ekeus, April 23, 1997.

36. R. Jeffrey Smith, "UN Pursuing 25 Germ Warheads," 1.

37. UNSCOM, S/1995/1038, Dec. 17, 1995, 24–26; S/1996/258, April 11, 1996, 21–22.

38. *Gulf War Air Power Survey (GWAPS)* (Washington: Office of the Secretary of the Air Force, 1993), vol. 2, pt. 2, 316, 328. Albright and Kelley, "Has Iraq Come Clean," 62. Smith, "UN Pursuing 25 Germ Warheads," 1.

39. Bush quoted in David Albright and Mark Hibbs, "Iraq's Nuclear Hide-and-Seek," *Bulletin of Atomic Scientists,* Sept. 1991, 14. *GWAPS,* vol. 2, pt. 2, 329. William S. Cohen, "In the Age of Terror Weapons," *Washington Post,* Nov. 26, 1997, A19.

40. *Plan for the Implementation of Relevant Parts of Section C of Security Council Resolution 687* (1991), S/22614, May 17, 1991, 2.

41. John M. Goshko, " 'Standoff Is Over'; U.N. Team Leaves Parking Lot,"

Washington Post, Sept. 28, 1991, A1; Marilyn Greene, "U.N. Team Leader: Grit, Good Humor during Tense Times," *USA Today,* Sept. 26, 1991, 2A.

42. First consolidated IAEA report, S/1996/261, April 11, 1996, 11. Albright and Kelley, "Has Iraq Come Clean," 63. Ambassador Rolf Ekeus, Speech to Nuclear Non-Proliferation Conference at the Carnegie Endowment for International Peace, Washington, June 10, 1997.

43. UNSCOM, S/1994/750, June 24, 1994, 5; S/1996/848, Oct. 11, 1996, 6; S/1995/864, Oct. 11, 1995, 18; S/1996/258, April 11, 1996, 16.

44. Central Intelligence Agency, *The Weapons Proliferation Threat* (Washington, March 1995), 13. Paul Richter, "Iraq's Toxic Cache Sizable, U.S. Believes," *Los Angeles Times,* Nov. 19, 1997. "Iraq Makes Lethal Gas in Sudan," *Sunday Times,* Nov. 16, 1997. "Iraqi Guards Die in Chemical Weapons Accident," *Electronic Telegraph,* Nov. 16, 1997.

45. UNSCOM, S/1996/848, Oct. 11, 1996, 22. Interview with UNSCOM official, March 18, 1998. UNSCOM, S/1995/864, Oct. 11, 1995, 7–8; S/1996/848, Oct. 11, 1996, 23; S/1995/284, April 10, 1995, 17; S/1995/284, April 10, 1995, 17–19. Richter, "Iraq's Toxic Cache."

46. Speech by Rolf Ekeus at the Washington Institute for Near East Policy, Jan. 29, 1997. Smith, "UN Pursuing 25 Germ Warheads," 1.

47. Smith, "UN Pursuing 25 Germ Warheads," 1. Testimony of Joseph Nye, U.S. Congress, Senate, Committee on Foreign Relations, *U.S. Policy toward Iran and Iraq,* 104th Cong., 1st sess., March 2 and Aug. 3, 1995, 173.

48. UNSCOM, S/1995/864, Oct. 11, 1995, 11; S/1996/848, Oct. 11, 1996, 14–16; S/1997/301, April 11, 1997, 24.

49. Interviews with UNSCOM officials, Nov. 20, 1997.

50. Speech by Ekeus, Jan. 29, 1997. UNSCOM official speaking off the record, Nov. 20, 1997.

51. R. Jeffrey Smith, "Did Russia Sell Iraq Germ Warfare Equipment?" *Washington Post,* Feb. 12, 1998, A1. Interviews with UNSCOM officials, March 1998.

52. Smith, "UN Pursuing 25 Germ Warheads," 1. Brad Roberts, email communication with author.

53. Aziz quoted in Nelan, "America the Vulnerable," *Time,* Nov. 24, 1997.

8. What Is to Be Done?

1. Scott Sagan argues that the U.S. government should explicitly rule out the use of nuclear weapons in response to chemical and biological attacks. I see no utility in being explicit.

2. PDD-39 U.S. Policy on Counter-terrorism, declassified and redacted version, from www.fas.org/irp/offdocs/pdd39.htm.

3. Interview with Alex Riedy, head of the U.S. team at Ulba, Aug. 14, 1997; Briefing Book for the Secretary of Energy for the Project Sapphire Press Conference. William Potter, "The Sapphire File: Lessons for International Nonproliferation Cooperation," *Transitions,* Nov. 17, 1995. John Tirpak, "Project Sapphire," *Air Force Magazine,* Aug. 1995.

4. "Operation Sapphire; Sapphire Team Accomplishments," Briefing Book for the Secretary of Energy.

5. Ibid.

6. Interview with a State Department official who interviewed Russian physicists, May 27, 1997.

7. Tirpak, "Project Sapphire," 53.

8. On the broader effort to secure fissile materials in Russia and the United States see Matthew Bunn and John P. Holdren, "Managing Military Uranium and Plutonium in the United States and the Former Soviet Union," *Annual Review of Energy and the Environment* 22 (1997): 403–486.

9. See Matthew Bunn, "Security for Weapons-Usable Nuclear Materials: Expanding International Cooperation, Strengthening International Standards," in *Comparative Analysis of Approaches to Protection of Fissile Materials: Proceedings of a Workshop at Stanford, California, July 28–30, 1997.* Livermore, Calif.: Lawrence Livermore National Laboratory, Document Conf.-97-0721, 1998.

10. Interview with U.S. government official who asked not to be identified, May 27, 1997. Letter from a senior Latvian official, Dec. 12, 1996.

11. Interview with Customs officials, May 28, 1997.

12. Information supplied by DOE, July 1, 1997.

13. "Overview of Initiative for Proliferation Prevention," manuscript, U.S. Department of State.

14. Mirage Gold Full Field Exercise, 10/16–10/21/94, New Orleans, FBI After-Action Report (redacted copy). Exercise Mirage Gold After-Action Report, March 1995, prepared by the Federal Emergency Management Agency. Both in U.S. Congress, Senate, Committee on Governmental Affairs, Permanent Subcommittee on Investigations, *Hearings on Global Proliferation of Weapons of Mass Destruction,* pt. 3, 105th Cong., 2nd sess., pt. 2, March 27, 1996.

15. Briefing slides from Defense Nuclear Agency regarding Mirage Gold and Mirrored Image Exercises, undated. In files of U.S. Congress, Senate, Committee on Governmental Affairs, Permanent Subcommittee on Investigations, in connection with *Hearings on Global Proliferation of WMD,* pt. 3.

16. "The Mile Shakedown Series of Exercises, A Compilation of Comments and Critiques," Feb. 18, 1995, 53; ibid., page no. removed, probably 42. Memorandum to Manager, Nevada Operations Office, DOE, prepared by Charles J. Beers Jr., Jan. 25, 1995, in U.S. Senate, *Hearings on Global Proliferation of WMD,* pt. 3, 141–143. DOE, NEST Assessment Team Report, July 12, 1995 (henceforth the Sewell report), 12, 18.

17. "Navigating the Course of Emergency Preparedness in New York City, An Analysis of Chemical Disaster Preparedness, Response and Planning, Based on the New York City No-Notice Mobilization Exercise, April 11, 1995" (redacted copy). Interview with FEMA officials, Feb. 18, 1997. Interview with Dr. Frank E. Young, former director of the National Disaster Medical System, July 16, 1996. Jonathan B. Tucker, "National Health and Medical Services Response to Incident of Chemical and Biological Terrorism," *JAMA* 278, no. 5 (Aug. 6, 1997), 362.

18. U.S. Senate, *Hearings on Global Proliferation of WMD,* pt. 3, 103.

19. Stephen Green, "Secretive Unit Awaits Call to Battle Nuclear Terrorism," *San Diego Union-Tribune,* Dec. 12, 1993, A1. Sewell report, 6–7.

20. Sewell report, 22; statement of Sarah Mullen, March 13, 1996; John Sopko and Alan Edelman, Staff Statement, U.S. Senate, *Hearings on Global Proliferation of WMD,* 104th Cong., 1st sess., pt. 1, Oct. 31, 1995. Interviews with DOE officials, 1996–1997.

21. Interview with Mike Austin of FEMA, Feb. 18, 1997.

22. Gregory Koblentz's interview with Gary Eifried, EAI Corporation, Feb. 28, 1997.

23. Barbara Starr, "At Hand to Deal with an Underhand Attack," *Jane's Defence Weekly,* Aug. 14, 1996, 17–19.

24. Interview with a senior official of the Office of Emergency Preparedness, June 3, 1997.

25. Ibid.

26. The Pentagon is also conducting "distant learning" to train the National Guard to respond to chemical and biological attacks. Defense Department Background Briefing Regarding the Inter-Agency Program Responding to the Nunn-Lugar-Domenici Legislation in FY 1997 Defense Authorization, April 16, 1997. The International Association of Fire Chiefs has estimated that equipping the 120 largest cities with ten sets of protection and detection gear would cost only $4.5 million. Memorandum from International Association of Fire Chiefs, from files of U.S. Congress, Senate, Committee on Governmental Affairs, Permanent Subcommittee on Investigations, in connection with *Hearings on Global Proliferation of WMD,* pt. 3.

27. Interviews with DOE personnel.

28. "Marines 1st to Help, But Not in U.S.," *Daily News* (New York), July 1, 1996. Defense Department Background Briefing regarding the Inter-Agency Program Responding to the Nunn-Lugar-Domenici Legislation in FY 1997 Defense Authorization, April 16, 1997.

29. Joint Report to Congress, "Preparedness and Response to a Nuclear, Radiological, Biological or Chemical Terrorist Attack," prepared by the Department of Defense and Department of Energy, June 1996, 11. Tucker, "National Health and Medical Services Response," 363.

30. Tucker, "National Health and Medical Services Response," 363.

31. Industry is also working on improving PCR detectors, but tends to focus on diseases that affect many people, such as HIV and influenza. Interviews with DOE scientists.

32. Presentation by G. W. Long et al., Biological Defense Research Program, Naval Medical Research Institute, at the International Conference on Emerging Infectious Diseases, Atlanta, March 1998.

33. Information provided by Lee Buchanan of DARPA. Interview with General Russ Zajtchuk, U.S. Army Medical and Materiel Command, Jan. 17, 1997.

34. The biologist Eileen Choffnes claims that because canaries on a chip cannot distinguish different organisms within a class, they may result in many false positives. Nonetheless they may be a useful warning signal. Interviews with Choffnes, March 1998.

35. Douglas Lowrie, "DNA Vaccination Exploits Normal Biology," *Nature Medicine* 4, no. 2 (Feb. 1998). Joan Stephenson, "Pentagon Funded Research Takes Aim at Agents of Biological Warfare," *JAMA* 278, no. 5 (Aug. 6, 1997); information provided by Lee Buchanan of DARPA.

36. Stephenson, "Pentagon Funded Research."

37. Information provided by DOE.

38. Office of the Attorney General, "The Attorney General's Guidelines on General Crimes, Racketeering Enterprise and Domestic Security/Terrorism Investigations." Interview with Robert Blitzer, FBI, Jan. 26, 1998.

39. Before the Oklahoma City bombing the FBI typically encountered about a dozen incidents per year involving threats, boasts, or actual attempts to acquire or use weapons of mass destruction; now the FBI handles about a hundred per year. Blitzer interview.

40. A former CIA director speaking off the record, June 6, 1997.

41. Clifford Krauss, "8 Countries Join in an Effort to Catch Computer Criminals," *New York Times,* Dec. 11, 1997, 12. Kenneth W. Dam and Herb Lin, eds., *Cryptography's Role in the Information Society* (Washington: National Academy Press, 1996), 294.

42. Interview with William Pierce, April 22, 1997. Former CIA director, June 6, 1997. John Deutch, "Think Again: Terrorism," *Foreign Policy,* Fall 1997.

43. Dam and Lin, eds., *Cryptography's Role,* 330. Deutch, "Think Again," 20.

44. Former CIA director, June 6, 1997.

45. U.S. Department of Justice, "Report on the Availability of Bomb-Making Information, the Extent to Which Its Dissemination Is Controlled by Federal Law, and the Extent to Which Such Dissemination May Be Subject to Regulation Consistent with the First Amendment to the United States Constitution," submitted to the U.S. House of Representatives and the U.S. Senate, April 1997, 13.

46. David G. Savage, "Did Hired Killer Go by the Book?" *Los Angeles Times,* May 7, 1997. *Rice v. Paladin Enterprises, Inc.,* 940 F. Supp. 836 (D. Md. 1996), appeal docketed, no. 96–2412 (4th Circuit). "Report on the Availability of Bomb-

Making Information." David Montgomery, "If Books Could Kill," *Washington Post,* July 26, 1998, F1.

47. 18 USC section 373.

48. Ibid.

49. 18 USC section 2. Section 323 of the Antiterrorism and Effective Death Penalty Act of 1996 (AEDPA), to be codified as amended in section 2339A(a) of Title 18.

50. 18 USC section 231(a)(1).

51. "Report on the Availability of Bomb-Making Information," 51.

52. Marie Isabelle Chevrier, "The Threat That Won't Disperse: Why Biological Weapons Have Taken Center Stage," *Washington Post,* Dec. 21, 1997, C1.

53. The former Soviet Union still has not opened its military biological weapons facilities to inspection. Ken Alibek, "Russia's Deadly Expertise," *New York Times,* March 27, 1998, A23.

54. See Robert P. Kadlec, Allan P. Zelicoff, and Ann M. Vrtis, "Biological Weapons Control: Prospects and Implications for the Future," *JAMA* 278, no. 5 (Aug. 6, 1997).

55. Interview with Larry Wayne Harris, Jan. 23, 1998. U.S. Department of Health and Human Services, Centers for Disease Control and Prevention, "Additional Requirement for Facilities Transferring or Receiving Select Agents: Final Rule," *Federal Register,* Oct. 24, 1996, 61: 55190, cited in James R. Ferguson, "Biological Weapons and U.S. Law," *JAMA* 278, no. 5 (Aug. 6, 1997).

56. President William Jefferson Clinton, Remarks at the United States Naval Academy Commencement, Annapolis, May 22, 1998.

57. Matthew Bunn, Kenneth Luongo, et al., "The Nuclear Weapons Complexes: Meeting the Conversion Challenge—A Proposal for Expanded Action," manuscript, Russian-American Nuclear Security Advisory Council, Princeton, N.J., Sept. 1998.

Acknowledgments

Many people helped me as I was writing this book. My dissertation advisors at the Kennedy School of Government at Harvard University, Graham Allison, Ash Carter, Paul Doty, Matthew Meselson, and Richard Zeckhauser, taught me about commitment and persistence as well as about weapons of mass destruction. I learned a lot from colleagues during my postdoctoral fellowship at Lawrence Livermore National Laboratory, notably Charles Ball, Debbie Ball, Sybil Francis, and Peter Lavoy. My colleagues at the National Security Council, particularly John Beyrle, Chip Blacker, Matthew Bunn, Nicholas Burns, Rosemarie Forsythe, Rose Gottemoeller, Sheila Heslin, Carlos Pasqual, Steve Pifer, and Frank von Hippel, taught me about how policy is made.

I have been extraordinarily fortunate in having five outstanding research assistants: Darcy Bender, Stacy Gunther, Greg Koblentz, Melinda Lamont-Havers, and Lynn Patyk. Joe Cirrincioni of the Carnegie Institution generously donated some of Stacy and Melinda's time. Many colleagues provided comments on drafts of the book; I am especially grateful to Robert Art, John Ausink, Steve Black, Polly Carter, Eileen Choffnes, Jerry Dzakowic, Lynn Eden, Steve Fetter, Walter Laqueur, Lore Leavitt, Steven Lee, Brad Roberts, Doug Stephens, Jonathan Tucker, Al Zelikoff, and Peter Zimmerman. I am also grateful for the assistance provided by my editors: Michael Aronson, Camille Smith, and Teresa Lawson. Sean Lynn Jones and Sam Barton helped enormously by suggesting titles for the book.

I would like to thank my husband, Jeff Frankel, for his love, patience, and understanding. And I could not have succeeded in any of these endeavors without the encouragement of my parents and the rest of my wonderful family.

My research was supported by a MacArthur Research and Writing Grant, a Hoover National Fellowship, the W. Alton Jones Foundation, and the U.S. Institute of Peace. Les Gelb and the Council on Foreign Relations made it possible for me to work at the National Security Council and awarded me a Next Generation Fellowship while I was finishing this book. I am immensely grateful for this support.

Index